Safety of the Nuclear Fuel Cycle

Edited by K. Ebert and R. v. Ammon

VCH

This book is a volume in the series
Nukleare Entsorgung – Nuclear Fuel Cycle
edited by Franz Baumgärtner, Klaus Ebert,
and Karl Heinrich Lieser

Volume 1, 1981 (in German)

Volume 2, 1983

Volume 3, 1986 (in German)

Volume 4, 1989 (present volume)

Safety of the Nuclear Fuel Cycle

Edited by Klaus Ebert and Reinhard v. Ammon

VCH

Prof. Dr. K. Ebert
Prof. Dr. R. v. Ammon
Kernforschungszentrum Karlsruhe
Institut für Heiße Chemie
P. O. B. 3640
D-7500 Karlsruhe 1

Published jointly by
VCH Verlagsgesellschaft, Weinheim (Federal Republic of Germany)
VCH Publishers, New York, NY (USA)

Printing: G. Gehringer GmbH, D-6750 Kaiserslautern
Bookbinding: Josef Spinner, Großbuchbinderei GmbH, D-7583 Ottersweier

Library of Congress Card No. applied for.

British Library Cataloguing-in-Publication Data:
Ebert, Klaus
Safety of the nuclear fuel cycle.
1. Nuclear power. Safety aspects
I. Title II. Series
621.48'35
ISBN 3-527-27806-0

Deutsche Bibliothek Cataloguing-in-Publication Data:
Safety of the nuclear fuel cycle / ed. by Klaus Ebert and
Reinhard v. Ammon. – Weinheim ; Basel (Switzerland) ;
Cambridge ; New York, NY : VCH, 1989
(Nuclear fuel cycle ; Vol. 4. 1989)
ISBN 3-527-27806-0
NE: Ebert, Klaus [Hrsg.]

Distribution:
VCH Verlagsgesellschaft, P.O. Box 101161, D-6940 Weinheim (Federal Republic of Germany)
Switzerland: VCH Verlags-AG, P.O. Box, CH-4020 Basel (Switzerland)
Great Britain and Ireland: VCH Publishers (UK) Ltd., 8 Wellington Court, Wellington Street, Cambridge CB1 1HW (Great Britain)
USA and Canada: VCH Publishers, Suite 909, 220 East 23rd Street, New York, NY 10010-4606 (USA)

Preface

For many years the safety of the nuclear fuel cycle has been a very sensitive issue within public discussion on nuclear energy production in Germany. Especially after the Tschernobyl accident and the demonstrations at the Wackersdorf reprocessing plant (now under construction) it became obvious that public acceptance of nuclear energy had severely diminished. The main cause is a deep distrust in its safety. This is also true for safety aspects connected with reprocessing and waste handling and its storage.
Therefore it was logical to adopt this topic in a seminar for chemists, physicists and engineers in the field of reprocessing and waste handling. The aim of the seminar was to discuss within a relatively small group of experts as thoroughly as possible and in a free and open manner safety related subjects of the nuclear fuel cycle. Some of these points are often addressed in discussions with opponents of nuclear energy.
This seminar took place in March 1987 in Obertraun (Austria) and was organized by the Institut für Heiße Chemie of the Kernforschungszentrum Karlsruhe (KfK) in cooperation with the Nuklearchemie division of the Gesellschaft Deutscher Chemiker (GDCh) and the Verein Österreichischer Chemiker (VöCh). Whereas previous seminars of this kind had focussed on various chemical and technical aspects of reprocessing nuclear fuel (see vol. 2 of this series), this time waste treatment and final storage were also included.
Twenty-four out of the forty-two papers presented at the seminar are contained in this volume. They give a good profile of the seminar, although reprocessing is somewhat overemphasized.
We would like to thank the contributors to this volume and to Dr. W. G. Filby who performed stylistic finishing on some of the manuscripts. Veronika Zwick deserves special acknowledgement for her efforts in preparing the final manuscript.

Karlsruhe,
in October 1988

Klaus Ebert
Reinhard von Ammon

Contents

1. Reprocessing

2. Waste Treatment and Storage

Part I
Reprocessing

Safety of reprocessing under the research and development aspect

H. Schmieder, K. Ebert

SUMMARY

Reprocessing is a technology which has been proven for more than 30 years and successfully demonstrated also for LWR fuels having undergone high burnups. The higher discharges of radioactive materials with the liquid effluents in the initial phase have been clearly diminished also in plants based on the seaside. It is a requirement imposed by the site that the limit values of the German reprocessing plant WAK are lower by several orders of magnitude. A further reduction in H-3 discharges can be expected from the H-3 scrub in the first extraction cycle. Two techniques for Kr-85 retention developed up to "cold" pilot operation have been used in order to reduce the level of discharges with the effluent air.

The causes of incidents have been recognized and largely understood so that the required protection can be taken into account in engineered safeguards. Engineered safeguards according ot the status of the art allow to safely comply with the requirements of the Radiation Protection Ordinance also under accident conditions.

The following R&D work is of importance, especially with a view to fuels with a higher Pu content:

1 - Techniques for separating wrong phases, especially the organic phase,
2 - Apparatuses offering a higher space-time yield,
3 - In-line concentration monitors,
4 - Computer aided process control,
5 - Applicability of new materials and, last but not least:
6 - Upgrading of remote handling technology.

Dr. H. Schmieder, Prof. Dr. K. Ebert, Institut für Heisse Chemie, Kernforschungszentrum Karlsruhe, Postfach 36 40, D-7500 Karlsruhe 1, Federal Republic of Germany

1 INTRODUCTION

Reprocessing operations have been conducted on an industrial scale for more than 30 years. The separation technique which has become generally accepted is liquid-liquid extraction with tributyl phosphate as the solvent and nitric acid as aqueous phase, the so-called PUREX process.

Worldwide more than 100,000 te of fuel in total have been processed so far by this technique. However, the major portion are fuels with a low burnup rate which are processed to obtain weapons grade plutonium. More than 30,000 te of metallic fuel have been processed until now from power reactors (mainly MAGNOX reactors). The amounts of nuclear fuels from light water reactors which had been processed until mid-1986 are shown in Fig. 1; they are equivalent to approximately 2500 te.

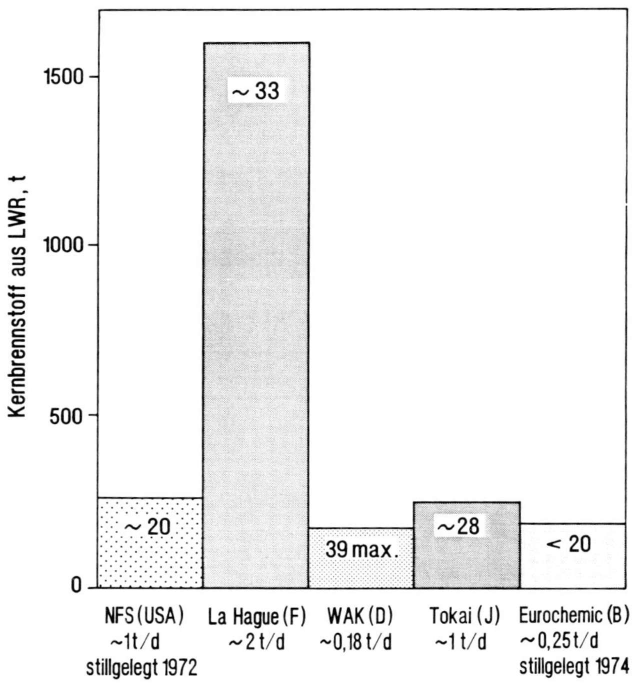

Fig. 1: Amounts of nuclear fuel from LWR reprocessed until mid-1986 (burnup in GWd/d)

The requirements made on the process have become more stringent since the early days of its application because the levels of burnup have been raised; and, on the other hand, fuels from power reactors are normally allowed to decay longer before they are processed so that more exacting requirements have not to be made, at least not as regards radiation exposure. During the last 10 years the efforts of the operators as well as development work have been concentrated on increasing plant availability, decreasing radioactive discharges, reducing the waste volume, and lowering the radiation exposure of the operating staff. Above all for nuclear fuels from future reactors which will undergo still higher burnups and contain more plutonium, it will be crucial to avoid a further rise in the costs of reprocessing. Therefore, long-term R&D activities are geared to simplifying the process /1, 2/. A major part of current R&D programs in the Federal Republic of Germany are devoted still now to upgrading the techniques of radioactive materials retention in the offgas, to reducing the waste volume, and to remote handling technology.

This paper serves to describe the level of safety achieved in reprocessing and to discuss related R&D work.

Questions of nuclear safeguards and fissile material security will not be treated here.

2 PURPOSES OF PROTECTION AND SAFETY CONCEPT

According ot W. Thomas /3/ the purposes of protection derived from the Atomic Energy Act and the Radiation Protection Ordinance can be summarized as follows in terms of reprocessing:

- Containment and shielding of the radioactive materials

- Minimization and monitoring of radioactive material discharge

- Minimization and monitoring of radiation exposure of the operating staff.

It has to be ensured by means of the engineered safeguards that these goals can be attained. The purposes of protection mentioned before are basically similar to those of a nuclear power plant. Nevertheless, the engineered safeguards differ greatly. Whereas in a nuclear power plant dangerous situations may develop quickly, this does not apply in the chemical reprocessing plant where these situations in most cases develop but slowly and can be monitored and recognized, e. g. by usual chemical analysis including sampling. This is why the requirements made on the reliability of instrumentation and control technology have not to be as exacting as in a nuclear power plant (operating automatically, redundant, diverse). If, for instance, the accident initiated by self-heating through the decay heat in a Nuclear Power Plant (NPP) is compared with that in reprocessing as regards the sequence and consequences, this difference becomes evident (Fig. 2).

	NPP (1.3 GWe)	Reprocessing (~ 50 GWe a*)
Radioactive Inventory, GCi	~ 10	⩽ 5 (<5 % liquid)
Fission products	Predominantly short-lived	Predominantly long-lived
Plutonium, te	~ 1	~ 20
Decay heat		
Power density, $MJH^{-1}I^{-1}$	~20 max (core)	~0.007** (HLWC)
Self-heating without cooling	Very quick	Slow
Most serious accident	Core melt	Boiling (after several days)

* equivalent amount of nuclear fuel inclusive of fuel element storage facility for 2 te/d reprocessing plant

** 7a decay time

Fig. 2: Comparison of NPP with reprocessing

Whereas the power densities of the radioactive inventories in the nuclear power plant and in the reprocessing plant are of the same order of magnitude, the power density in the reactor core after the chain reaction has stopped is higher by more than the factor 1000 compared with that in the high level liquid waste concentrate (HLLWC) of the reprocessing plant. In case of total failure of the coolant system, a critical condition would develop in the reactor core within minutes. The most serious accident here would result from a core melt.

In the HLLWC vessel the solution would be brought to boiling after some days only /4/ when the cooling system fails. Besides, rise in concentration is avoided by a passive cooling system installed in addition and working on the evaporative cooling principle /5/. As far as one can judge, the most serious accident in this example would therefore be caused by boiling. However, boiling leads to only a slight increase in concentration of radioactive materials in the offgas which are largely trapped by the retention systems.

The safety concept of a reprocessing plant relies on the so-called barriers. Basically, the environment of the plant is protected by three barriers:

1. Process equipment (apparatuses, vessels, pipework, etc.)

2. Process cells with linings and pans

3. Building

The necessary inlets and outlets of the plant for the starting product and the end product, respectively, as well as the process media (air, water, etc.) are protected by retention systems. Each of the three barriers is equipped with retention systems for gaseous, liquid and also solid radioactive materials, depending on the potential level of radioactivity, in order to avoid discharges into the environment. These retention systems include offgas cleaning trains, filters, ion exchangers, buffer vessels, etc. A major element of the safety concept consists in checking the integrity of the barriers and the functioning of the retention systems. This is achieved by instrumentation or sampling. Moreover, the retention systems must be designed in such a way that also leakages at the barriers do not result in inadmissible discharges. Taking into account practical experience, the required design data must be worked out in accident studies. The essential variables which can be influenced and which govern radioactive discharges in accidents include the radioactive inventory involved, the actual conditions of transport (pressure, flow, temperature), and the effectivity of the retention systems under these conditions. Within the framework of the "Safety Studies on Nuclear Waste Management" Project this type of accident study was performed /6/. According to the assumed plant design, the discharges of radioactive materials under conditions of "serious" accidents are not inadmissibly high.

3 EXPERIENCE GATHERED IN OPERATION ACCORDING TO SPECIFICATION

3.1 RADIOACTIVITY DISCHARGES INTO THE ENVIRONMENT

The values of radioactive material discharges with the liquid effluents and the effluent air have been published for most of the facilities.

From the British facility at Sellafield larger amounts of radioactive materials have been discharged into the sea than with the liquid effluents from the french facilities at La Hague. Conditioned by the site of the Karlsruhe Nuclear Research Center, the radioactivity discharges with the liquid effluents, including those from the WAK reprocessing plant, are lower by five orders of magnitude, except for tritium. Figure 3 shows a comparison of pipeline discharges at

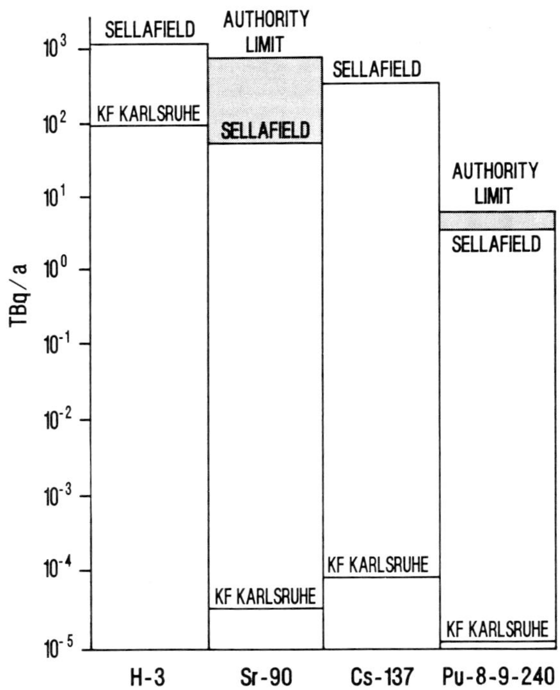

Fig. 3: Radioactive discharges with liquid effluents for KfK, including WAK, in 1985. Same representation for BNFL Sellafield (pipeline discharge) in 1985

Sellafield with those from the Karlsruhe Nuclear Research Center for tritium, Sr-90, Cs-137 and plutonium alpha-emitters, applicable for 1985. For alpha-emitters and Sr-90 the authority limits for Sellafield have been entered in adddition. These limits have been noticeably lowered during recent years /7/. Moreover, additional retention systems have been installed at Sellafield.

The discharges of radioactive materials with the effluent air have been represented in Fig. 4 where the Sellafield facility is compared with WAK. The emissions from WAK /9/ are higher in relative terms than the discharges with the liquid effluents from Sellafield. This may be explained by the differences of the processes applied in the two facilities /10/.

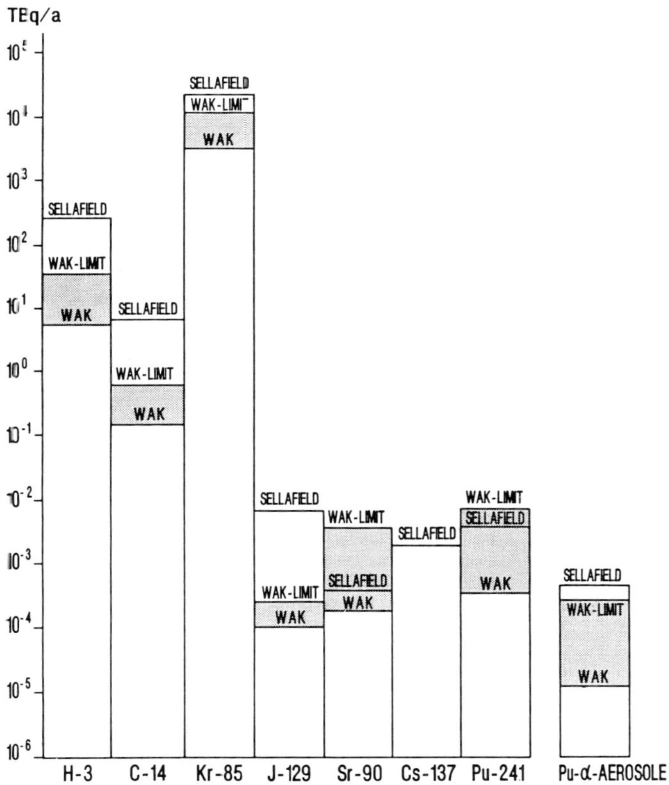

Fig. 4: Radioactive discharges with the effluent air for WAK together with authorized limits (1987) and for BNFL Sellafield (no limits specified) in 1985

The requirement of minimizing aqueous discharges usually does not allow the same extent of minimization on the gas path. However, on account of the large volume of aqueous discharges admissible at Sellafield, relatively low discharges with the effluent air can eventually be achieved more easily without necessitating much expenditure in process engineering.

For Kr-85 retention two techniques have been developed up to the stage of "cold" pilot operation /11/; if needed, they can be brought to technical maturity. No proven process is available now which would allow a noticeable reduction in aerosol emissions; this is true also for the actually low I-129 releases /8, 23/.

For H-3 lower releases in relative terms should be expected from the optimization of water and acid recycling provided that process engineering provides for H-3 scrub in the first cycle. But on account of the complex process involved and its interlinking, the actual reduction in H-3 releases can be determined solely in practical plant operation.

3.2 EXPOSURE TO RADIATION

An international comparison of the exposure to external radiation of the operating staff in shown in Fig. 5 /3/. Despite the fact that the rates of burnup and the throughputs are increasing, a clear tendency towards a lower radiation exposure can be recognized for the facilities still in service. Radiation exposure related to the throughput at LA HAGUE dropped from 5.8 Sv/GWa to 1.1 Sv/GWa from 1975 until 1983 /3/. A striking feature in Fig. 5 are the high values applicable to the NFS facility and the low values achieved at the Japanese plant at Tokai Mura. At NFS (decommissioned in 1973) only inadequate decontamination factors were achieved during water and acid recycling which was due to the process applied. It is reported that this had contributed much to radiation exposure. In general, major contributions to radiation exposure result from repair and analytical work. It can be expected that the application of remote handling techniques as well as improved automated sampling and analytical techniques will allow to achieve a further clear reduction in radiation exposure. In terms of process engineering, the Tokai Mura plant resembles the French plant and was built under supervision of a French engineering company. The extremely low values achieved as early as during initial operation are probably due rather to restrictive administrative measures than to causes to be found in plant technology.

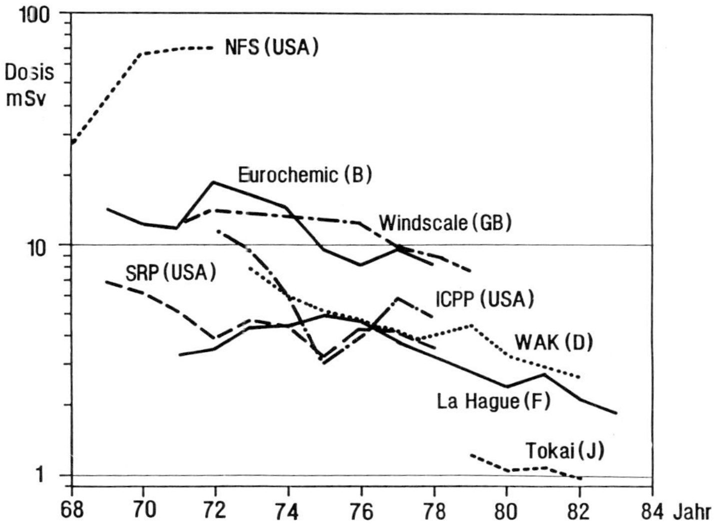

Fig. 5: Mean individual doses received in reprocessing plants according to W. Thomas /3/

4 EXPERIENCE ACCUMULATED DURING INCIDENTS

Quite a number of incidents are known from the literature /12, 7/ which have occurred in radiochemistry, fuel manufacture and reprocessing with various degrees of damage involved. They will be divided into four groups and discussed here:

- explosion
- fire
- criticality
- leakage.

In Fig. 6 the causes of these types of incidents and the measures taken to avoid them have been summarized. The causes of incidents due to external events which have been entered on the last line of the table will not be dealt with in more detail here because they were discussed elsewhere /15/.

Event	Causes	Countermeasures
Explosion	Explosive compounds (red-oil, azides?) Explosive gases	Avoiding both formation and critical conditions Dilution and avoiding ignition sources
Fire	Zircaloy Solvent (has not occurred) Cable insulation	Inertization Conventional fire defending system
Criticality	Accumulation not in conformity with specification Maloperations	Geometry, "poisoning," instrumentation Instrumentation, administration
Leakages	Technical failure Corrosion Maloperations	Emergency systems Instrumentation
External events	Earthquakes, plane crashes, etc.	Resistance of building and equipment (according to Gasteiger)

Fig. 6: Causes of failures and envisaged countermeasures

Explosive decompositions of solvent(DBP)-heavy metal compounds occurred repeatedly in the early days of reprocessing. These incidents have mostly been described under the heading "red oil" in the literature. The causes and sequences of them are well understood today /13/. By improved solvent separation from the evaporator feed solutions and by setting a limit on the temperature at < 135° C these incidents can be avoided. In Fig. 7 important events of this type have been compiled. The damage was high which had been caused by inadvertent feed of about 170 kg $UO_2(NO_3)_2 \cdot 2TBP$ into the denitrator of Savannah River in 1975 although no radiation exposure had occurred. A similar incident can be safely avoided by improvement of phase separation and installation of control instrumentation. The explosion which had the most serious impact occurred at Windscale (Sellafield) in 1973. The solvent (Butex) had accumulated in an uncontrolled manner in a vessel together with dissolver residues, had decomposed through explosion and heavily contaminated the working area in front of the head-end cell. This led to high individual doses by Ru-106 incorporation. The lesson to be learnt from this incident is the

Cause	Event	Damage
Rapid decomposition of organic material in evaporators Savannah River, USA, 1953	Explosive decomposition, probably by organic material, in evaporator containing HNO_3 and radioactive material.	Considerable damage to equipment, no personal injury.
Oak Ridge, USA, 1959	Explosive decomposition in evaporator (THOREX facility, shielded) during evaporation of HNO_3-Pu solution with decontaminant.	Considerable contamination with plutonium within building and outside; no discharge by stack.
Rapid decomposition of TBP in the denitrator Savannah River, USA, 1975	$UO_2(NO_3)_2 \cdot 2TBP$ (170 kg) fed inadvertently into the $UO_2(NO_3)$ denitrator. Fast thermal decomposition accompanied by pressure buildup (500 m^3 gas, 18 m^3/s; $T > 170$ °C) Release of energy $\sim 7 \times 10^5$ kcal.	Building damaged; no radiation exposure.
Rapid decomposition of Butex together with dissolver residues Windscale, GB, 1973	Accumulation not in conformity with specification of extractant (Butex) and dissolver residues in feed tank led to explosive decomposition accompanied by pressure buildup in the head-end cell and release in the working area.	Heavy radiation exposure, 1000 rem (50 a; Ru) at the maximum; heavy contamination, stack discharge: 1/5 of daily limit.

Fig. 7: Explosion events (selected by extent of damage and relevance to LWR reprocessing)

necessity of controlled separation of organic wrong phases. This is achieved today by decanters and vapor or kerosine scrubs.

A possible cause of explosions taking place during reprocessing can be seen to lie in the formation of azides. However, if NO_x is present in excess in the offgas system by which HN_3 is decomposed such an incident can be excluded /14/. In alkaline dissolution of aluminum used as the fuel clad, a process associated with the formation of NH_3, the silver reactor used for iodine retention exploded at Hanford in 1958. Although this caused considerable damage within the cell, there had been no contamination of the environment. No azide formation as the cause of explosion had been detected. Given the different type of process used in fuel dissolution, this incident does not bear any direct impact on LWR reprocessing.

The hazard of explosive gas mixtures being present is safely controlled today in reprocessing plants by dilution which keeps the values below the explosion limits and by avoiding sources of ignition. No incident is known which has resulted from such a cause.

The main causes of incidents by fires and the measures taken to avoid them have been compiled in Fig. 6. Figure 8 shows major incidents of this type.

A number of fires, especially in the USA, occurred when the highly pyrophorous metals plutonium and uranium were processed. Although Pu-quantities of the order of kilograms burnt, directed flow and aerosol filtering succeeded in keeping the releases to the environment at minimum levels. These incidents are not important in LWR reprocessing because U and Pu are present as oxides. This is true also for Magnox fuel reprocessing involving cladding materials made of magnesium alloys. No fire has been reported which would have been caused by Zircaloy chips or dust in LWR reprocessing. However, it can be derived from experiments that this might be a potential cause of fire in LWR reprocessing /16/. The inertization of the cell discussed as a precaution to avoid solvent fires will likewise reduce the risk of Zircaloy inflammation. Solvent fires have actually not become known which seems to be plausible if one takes into account the relevant physical properties and the operating conditions prevailing.

A serious damage which, however, did not give rise to inadmissible releases of radioactivity occurred at La Hague in 1980 when a cable

Cause	Event	Damage
Pu and U metal fires (most frequent cause of fire in the USA)	Not relevant to oxide fuel reprocessing. Mostly ignition of chips from machining of the two highly pyrophorous metals with burning of up to kg-amounts of solid Pu blocks (22 kg, in Rocky Flats, USA, 1957). Also in the presence of other burnable materials.	If staggered filter system intact, no or only minor release outside the controlled area.
Magnox-U fires Windscale, GB, 1979	Little relevant to oxide fuel reprocessing. Several grams of U metal fuel ignited in addition Magnox residues in highly shielded head-end.	Substantial release (controlled area and environment). Eight operators received $\leqslant$ 6 % of annual limit (Cs).
Cap de la Hague, F, 1981	Fire of Magnox residues in storage bin. Probably ignited by cotton used as a U-decontaminant.	Substantial release on ground (fence of premises), 370 Bq/m^2. Maximum individual dose 5.7 rem.
Cable fire Cap de la Hague, F, 1980	Fire at switching panel caused heavy cable fire extending to emergency current supply system. By mobile diesel generators HLLWC cooling could be restarted after 1 hour. Provisional supply of the total ventilation system after 14 hours.	Destruction of primary distribution switchgear, control systems for emergency current, failure of total energy supply system. HLLWC tank: from 40 °C to 44 °C (1 hour). Fuel element storage basin: ΔT 1-2 °C.

Fig. 8: Fire events

was set to fire causing the emergency current supply system to fail. By use of mobile diesel generators HLLWC vessel cooling could be restarted provisionally after one hour and the ventilation system after 14 hours. It should be possible to exclude such an incident safely by redundancy of current supply systems.

From our point of view two questions are important in the context of explosion and fire prevention:

- Deepening the knowledge of Zircaloy inflammation
- Study of the explosive TBP-heavy metal compounds and determination of the critical temperatures.

Investigations relating to these questions are already under way. Moreover, it should be examined once more whether use of resins for dissolved TBP absorption will not, eventually, be more advantageous, at least for some few evaporator feed solutions.

Eight criticality incidents have become known which occurred in reprocessing plants or in similar facilities where chemical nuclear processes are performed /3/. All of these incidents concerned solutions and were caused by accumulations not conforming to specification or by maloperations. Five of these incidents have been summarized in Fig. 9; three of them happened at Idaho where highly enriched uranium (90 % U-235) is being reprocessed. The U-235 equivalent in the fuel solution of LWR reprocessing plants is < 5 %. The highest amount of energy was released in an incident which took place in 1959; 4×10^{19} fissions occurred over several hours. As in most of critical configurations in solution, the first pulse produced about 10^{17} fissions (750 kcal). It is important to note that the energy in these configurations is released relatively slowly so that a comparison is not possible with a conventional explosive or even a nuclear explosive as regards the hazard involved /17, 18/. As the solution gets distributed in the vessel or equipment, the critical configuration becomes quickly subcritical and after recombination of the flows it can become critical again (pulsed process). In an incident at Los Alamos an operator received a lethal radiation dose. However, he worked without any shielding. If the usual concrete shieldings are provided, the radiation doses to be expected are low unless fission products escape into the working area. Above all the two incidents at Los Alamos in 1958 and at Windscale (Sellafield) in 1970, when criticality ultimately occurred through an organic phase

Location	Fissile Material Configuration	Event	Energy Release (at slow rate into solution) Σ fission	kcal total	kcal 1. pulse	Damage
Idaho Chemical Process. Plant, USA (1959)	91 % U-235; 34 kg in 800 l inorganic solution	Filling not in conformity with specification of a container with critical geometry (19,000 l) behind concrete shielding	$\sim 4 \times 10^{19}$	$\sim 3 \times 10^{5}$	~750	Radiation exposure due to contamination (≲50 R, skin)
Los Alamos Scientific Laboratory, USA (1958)	PU, 3.27 kg in 160 l organic solution	Long-term accumulation not in conformity with specification of stirring vessel with critical geometry (850 l) accompanied by Pu enrichment in organic layer of solution (20 cm). No shielding!	$\sim 1.5 \times 10^{17}$	~ 1150	(~ 1150)	Extremely high radiation exposure of an operator ($12 \pm 6 \times 10^{3}$ R). No fission product contamination; plant not destroyed
Idaho Chemical Process. Plant, USA (1961)	Enriched U, 8 kg U-235 in 40 l inorganic solution	Escape of solution from evaporator with subcritical geometry into condenser with critical geometry, behind concrete shielding	6×10^{17}	~ 4600	(~4600)	Radiation exposure ≲55 rem. No FP contamination; plant operation resumed after 1 hour
Idaho Chemical Process. Plant, USA (1978)	Highly enriched U, about 22 g/l in extraction column	Column operation not in conformity with specification resulted in FP accumulation	2.7×10^{17}	~ 2500		No radiation exposure; no contamination
Windscale, GB (1970)	PU, ~2.5 kg in 40 l organic solution	Accumulation of "heavy organic phase" in intermediate vessel behind concrete shiedling	10^{15}	~8		No major impacts

Fig. 9: Criticality events (eight in total occurred in reprocessing plants and similar facilities)

not detected, are relevant to LWR reprocessing plants. The requirement to be derived from these incidents, namely the completest possible separation of the organic phase, is in conformity with the requirement of avoiding explosions.

Thanks to the use made of an advantageous geometry, coupled with heterogeneous neutron absorbers and reliable fissile material monitors, adequate means are available today which practically rule out criticality incidents in LWR reprocessing. With respect to fuels more rich in fissile material (Pu) the following R&D tasks are important to simplifying the criticality concept:

- Equipment providing a higher space-time yield (fast contactors)

- Reliable in-line measurement of concentration (saving vessel volume)

- Computer aided process control (model improvement, model reduction).

A current development, which might be of importance to LWR reprocessing too, is a process in which by simple change of two parameters Pu accumulation /19/ in the extraction/fissile product scrub is avoided /20/.

The incidents termed leakages in Fig. 6 include damage to the barriers and retention systems which is caused essentially by technical failure, corrosion and maloperations. Emergency systems and redundancy in design as well as adequate instrumentation are measures which are indispensable to avoid these incidents.

Typical incidents have been compiled in Fig. 10. The greater number of incidents listed there were caused by maloperations and technical failure. Numerous incidents are attributable to the lack of monitors or were not detected in time because monitors were missing. In the majority of cases the consequences were minor. the most serious incident was a leakage occurring at Windscale (Sellafield) in 1979 when 10 m^3 of high level liquid waste perculated into the soil unnoticed. Failure of a liquid distributor and leakiness of the soil pan were the causes. The incident remained undiscovered for years because plant monitoring was inadequate.

Cause	Event	Damage
Technical failure, maloperations Numerous incidents	Uncontrolled discharge of pollutants by overflow, failure of valves, etc.	Mostly little
Windscale, GB, 1979	Malfunction of HLW liquid distributor led to tank overflow to leaking lining and bottom. Undiscovered for several years.	~10 m^3 HLW in soil (10^5 Ci). High dose of soil under building.
Savannah River, USA, 1961 (Windscale, GB, 1981)	Short-cooled fuel element processed by mistake. I-131 retention not ensured.	~150 Ci I-131 in the environment of the plant. Milk maximum: 160 Bq/l.
Failure of retention systems for gaseous substances Several incidents	Failure of gas scrubbers and filters (leakiness, destruction) led to release via stack.	Rather low. (Hanford, USA, 1954: 260 Ci Ru-106 released).
Corrosion Numerous incidents	Mostly within the cells at evaporators, dissolvers, etc.	Rarely any release. Major impairments of operation, sometimes long outages.
Hanford, USA, 1973	Leak at high-activity storage tank; about 400 m^3 perculated into soil (~30 m above ground water level).	About 4×10^4 Ci Cs-137 1.4×10^4 Ci Sr-90, 4 Ci Pu-239 Persons and Columbia River not contaminated.
Windscale, GB, 1975	Via leak at storage basin Cs-137 and Co-60 contaminated liquid effluents into canals discharging into Calder River.	Contamination of river. < maximum admissible limit for drinking water.

Fig. 10: Leakage event

Two incidents at Savannah River and at Windscale (Sellafield) caused important releases of I-131 because short-cooled fuel elements were dissolved by mistake. This type of incident can now be safely avoided by installation of burnup monitors and by administrative precautions.

Some incidents have been reported from the early days of reprocessing when radioactive materials were released due to leakages in the offgas retention system or failure of gas scrubbers. With the exception of the incident at Hanford, USA, 1954, the releases were rather low.

The most serious incident caused by corrosion of storage tanks made from black steel occurred at Hanford in 1973. Approximately the same amount of activity perculated into the soil as at Windscale in 1979. Also at Windscale, contaminated liquid effluents escaping from a leak in a storage tank in 1975 contaminated the Calder River, although at a low level. Leakages of vessels with a potential risk of release to the environment are controlled today by providing a double containment and activity monitors and by use of upgraded materials.

Most of the incidents due to corrosion occurred at the process equipment, i. e. at the first barrier so that they resulted in only minor or no releases at all of radioactive materials. However, operation was greatly impaired. In Tokai Mura and at the WAK leaks at the dissolver and at the evaporator necessitated long interruptions of operation.

Regarding incidents due to leakages, the current R&D programs on corrosion of stainless steels as well as the examinations of special metals such as Ti, Zr, Hf, Ti-Ta alloys as well as Ta are important /21, 22/.

LITERATURE

/ 1/ W. L. Wilkinson
"International Meeting on Solvent Extraction and Ion Exchange in the Nuclear Fuel Cycle", AERE Harwell, 3-6 Sept. 1985

/ 2/ H. Schmieder, E. Henrich, K. Ebert
ENC'86, Geneva, June 1-6, 1986

/ 3/ W. Thomas
GRS-Fachgespräch 1984, Cologne, Gürzenich, 12 and 13 Nov. 1984

/ 4/ R. Gasteiger
in Wissenschaftliches Kolloquium, 12-14 May 1986, Evangelische Akademie Tutzing

/ 5/ F. Mayinger, H. Sandner, H. Völkl, E. Gelfort, B. Heits, I. Mischke
Atomkernenergie, Kerntechnik 49, 209 (1987)

/ 6/ Projekt Sicherheitsstudien der Entsorgung, January 1985, Hahn-Meitner-Institut für Kernforschung, 1000 Berlin 39 - Wannsee

/ 7/ Reports prepared by Gesellschaft für Reaktorsicherheit mbH, unpublished

/ 8/ H. Schüttelkopf, M. Pimpl
IAEA-SM-257/100, p. 395

/ 9/ W. Hoffmann, WAK, private communication

/10/ E. Henrich in F. Baumgärtner
"Chemie der Nuclearen Entsorgung", Teil III, Verlag Karl Thiemig, Munich 1980, p. 85

/11/ F. Weirich, this seminar

/12/ E. H. Schulz
"Vorkommnisse und Strahlenunfälle in kerntechnischen Anlagen", Verlag Karl Thiemig KG, Munich 1966

/13/ R. Becker, L. Stieglitz, H. Bautz, R. Will
Report KfK-3983 (1985)

/14/ D. Ertel, H. Schmieder, A. Stollenwerk, this seminar

/15/ R. Gasteiger, B. Heinrich
Proceed. Ann. Meeting on Nucl. Technol. 1987, Deutsches Atomform, Kerntechn. Gesellsch., Bonn 1987, p. 219 and p. 223

/16/ R. Würtz, this seminar

/17/ J. Döhler, E. Hörmann, T. Schulenberg, this seminar

/18/ B. Gmal, J. Weber, this seminar

/19/ J. Schön, M. Kluth, H.-J. Bleyl, H.-J. Schmidt this seminar

/20/ G. Petrich, this seminar

/21/ J. W. Schulze, this seminar

/22/ G. Marx, et al., this seminar

/23/ E. Henrich, K. Ebert
Proceed. 18th DOE Nuclear Airborne Waste Management and Air Cleaning Conf., CONF-840806 (M. W. First, Ed.), 1985, p. 780

Modelling of criticality accidents in aqueous solutions of a nuclear reprocessing plant

J. Döhler, E. Hörmann, T. Schulenberg

SUMMARY

The following results were obtained by applying a model to criticality excursions in nuclear reprocessing plants:

- The vessel geometry influences the nuclear power and the duration of criticality. Low liquid levels in large vessels produce higher nuclear power levels and criticalities of the order of a few minutes (nucleate boiling). High liquid levels in small vessels, however, produce low power and a duration of the accident of several hours (surface boiling).

- When liquid addition is interrupted in due time criticality terminates by itself by increasing the concentration up to the subcritical geometry.

- When the boiling temperature is reached criticality is generally not terminated by adding a solution free of fissile material but the number of fissions still increases.

- For the typical concentrations and quantities of fissile material in a nuclear reprocessing plant, it can be practically excluded that a criticality accident of more than 10^{20} fissions is produced.

J. Döhler, E. Hörmann, T. Schulenberg, Dornier System GmbH, Postfach 13 60, D-7990 Friedrichshafen 1, Federal Republic of Germany

1 INTRODUCTION

An important safety aspect in the design and operation of a nuclear fuel reprocessing plant is the protection against a criticality event. In a German nuclear reprocessing plant such an event is considered to be a design basis accident. This means that its radiological consequences must be estimated for licensing and must be kept below the limit given by the German Atomic Law. These estimates are usually derived from performed criticaltiy experiments such as the CRAC experiments, and from criticality excursions which have already occurred in solution of fissile material.

A description of the behaviour of criticality by means of models has so far only been possible for the first power maxima /1/.

The models of the initiation and long-term behaviour of a criticality accident in homogeneous, aqueous solutions of fissile materials presented here were part of the criticality accident analysis of the "Projekt Sicherheitsstudien Entsorgung (PSE)". This project aimed at developing methodical instruments for the performance of safety analyses for reprocessing plants /2/.

2 GENERAL BEHAVIOUR OF CRITICALITY ACCIDENTS IN AQUEOUS SOLUTIONS

Criticality in aqueous solutions of fissile materials shows a completely different behaviour from criticality in a heterogeneous fissile material/moderator system as it is, for instance, realized in a nuclear reactor.

At criticality in solutions of fissile materials the energy release is spread over the whole volume. The temperature in the solution can only rise to the boiling limit during the excursion. After heat-up the solution starts to evaporate which decreases the solution volume and increases the concentration of the fissile material. The mechanisms to limit the excursion, such as thermal expansion of the solution and formation of radiolytic gas and steam bubbles, have an almost undelayed effect. Thus, the aqueous solution of fissile material is provided with a strong self-limiting effect, especially in the initial nuclear power peak of a criticality excursion. Therefore, the energy release of the first power peak is by one or two orders of magnitude smaller than in heterogeneous fissile material/moderator systems. Therefore, criticality excursions in solutions of fissile materials will not have the same disastrous consequences as the SL-1 accident /3/ in the USA in 1961 or the Chernobyl accident /4/. Generally, the fission rate is stabilized after an initial power peak to a fairly constant power level as was confirmed by criticality accidents and experiments /5/.

2.1 KNOWN CRITICALITY ACCIDENTS

12 criticality accidents in solutions of fissile materials have become known so far /6, 7, 8/. Four of them took place in nuclear reactors and criticality experiments, eight in the chemical processing of uranium and plutonium.

The causes for the initiation were:

- drawing of control rods (in two experiments)
- transfer or filling of solutions of fissile material into critically unsafe vessels
- accumulation of fissile material in organic or aqueous solutions.

In eight accidents, the criticality excursion had a characteristic initial power peak. The maximum energy released in the first power peak amounted to 5.2 kWh or $6 \cdot 10^{17}$ fissions. The vessels remained intact in all accidents. From open vessels liquid was ejected. A nearly constant level of the nuclear power of about 20 min up to 37.5 h appeared in four accidents. The criticality accident with the maximum total number of $4 \cdot 10^{19}$ fissions /9/ took place in a comparably high solution volume of 800 l. The criticality was terminated by itself after the evaporation of 500 l of water. In only one accident in the uranium and plutonium chemistry the criticality was terminated manually /8/.

2.2 CRITICALITY EXPERIMENTS

A series of experiments was performed to analyse criticality excursions in aqueous solutions. Already in the late fifties the behaviour of aqueous homogeneous nuclear reactors was studied in the KEWB experiments /10/ in the USA. The CRAC experiments /5/ performed in France have so far been the most extensive studies to analyse criticaltiy excursions in queous solutions of fissilie materials. Tests in the boiling range were also carried out in these experiments. They were performed in two vessels of different size (300 mm and

800 mm 0). The uranium concentration (93 % U-235) was varied between 21 and 325 g/l and the volume flow rate between 97 and 1864 l/h. The time dependent fission rate, the integral fission number and temperatures were measured in all tests. Additional studies with respect to boiling behaviour and fission gas release are being performed with the SILENE experiment /11/ in France.

3 CONDITIONS FOR THE INITIATION OF CRITICALITY IN A NUCLEAR REPROCESSING PLANT

For a restriction of fissile material to an equivalent value of < 1.8 % U-235 before the solution flows into the extraction process, as it is planned for a future German LWR reprocessing plant, criticality cannot occur in pure uranium solutions. Criticality in such a plant is only possible in connection with plutonium.

3.1 PROTECTIVE MEASURES AGAINST CRITICALITY

In a future German LWR reprocessing plant, the following basic measures will be taken to avoid criticality /12/: The criticality safety is realized in the first extraction cycle by a limited concentration of fissile material. No special measures are required in the uranium purification and concentration cycles due to the restriction of fissile material to < 1.8 % U-235. The plutonium purification and concentration cycles are designed as geometrically safe against criticality. Before the solution is transferred to critically unsafe areas, the content of the fissile material is measured by a double analysis. Neutron monitors are installed on the storage vessels which give alarm when a limit is exceeded.

3.2 MODELLING OF THE INITIATION OF A CRITICALITY

The most frequent cause of criticality excursions has been the inadvertent transfer of fissile material into critically unsafe areas. In an advanced nuclear reprocessing plant, this transfer is very unlikely due to the provided safety precautions but it cannot be generally excluded. Therefore, the initiation of criticality in such an inadvertent transfer of plutonium nitrate solution is analysed in detail. A volume of fissile material becomes critical when the neutron multiplication factor exceeds 1. The infinite-medium neutron multiplication factor k of a typical plutonium nitrate solution in a nuclear reprocessing plant is < 1 for concentrations < 11 g/l Pu /13/.

Consequently, there must be a plutonium concentration < 11 g/l due to the finite geometry of the vessel for the initiation of criticality. This is only conceivable in the case of an inadvertent transfer from the plutonium pruification (20 g/l Pu) or the concentration cycles (200 g/l Pu) into geometrically unsafe areas.

The initiation of criticality is calculated as a function of vessel geometry, concentration of plutonium nitrate solution feed and solution already available in the vessel. The well known equation for the effective multiplication factor k_{eff} is /13/:

$$k_{eff} = \frac{k_{\infty}}{1 + B^2 \cdot M^2} \geqslant 1 \rightarrow \text{criticality}$$

M = migration length (material data of the solution of fissile material

B = buckling (geometric data of the vessel)

For cylinder geometry, the buckling is calculated as follows:

$$B = \left(\frac{2.4045}{R + \lambda}\right)^2 + \left(\frac{\pi}{h + 2 \cdot \lambda}\right)^2$$

R = vessel radius

h = filling height

λ = extrapolation distance

Fig. 1 demonstrates, as an example, the conditions for the initiation of criticality in the case of the inadvertent transfer of an aqueous solution with 20 g/l plutonium. Three standard sizes of vessels are considered. For empty vessels before the inadvertent transfer, the following minimum quantities of plutonium nitrate solution result in a criticality:

Vessel Diameter	Solution Quantity	Plutonium Quantity
110 cm	200 l	4 kg Pu
180 cm	550 l	11 kg Pu
280 cm	1400 l	28 kg Pu

If a plutonium-free solution is already available in a vessel, these plutonium quantities will increase. Similar considerations were made for plutonium concentration of 200 g/l.

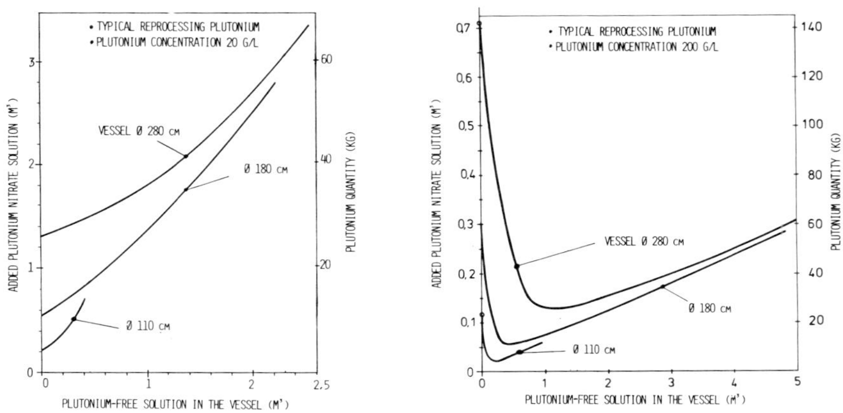

Fig. 1: Conditions for the initiation of criticality caused by an inadvertent transfer of plutonium solution into vessels of different sizes

3.3 MODELLING OF REACTIVITY INSERTION

The faster the effective neutron multiplication factor is increased in the critical state by adding fissile material, the high is the initial power peak, in general. The relative variation of the effective neutron multiplication factor k_{eff} in the critical state is called reactivity ρ.

$$\rho = \frac{k_{eff} - 1}{k_{eff}} \cdot 100 \qquad \text{in \%}$$

Fig. 2 demonstrates, as an example, the reactivity insertion versus time in the critical state at an inadvertent transfer of plutonium nitrate solution with a plutonium content of 20 and 200 g/l into vessels of different size. The filling rate is 1 l/s. A homogeneous mixture of the solution is assumed. The maximum reactivity insertion was calculated by using the critical equation to be 0.1 %/sec in the case of an inadvertent transfer of plutonium solution of 20 g/l into an empty vessel of 110 cm 0. Compared to this case, the reactivity insertion was considerably higher for the following experiments:

- CRAC experiments max. 0.6 %/sec. / 5/
- KEWB experiments max. 4.0 %/sec. /10/
- Criticality accidents max. 60 %/sec. / 6/

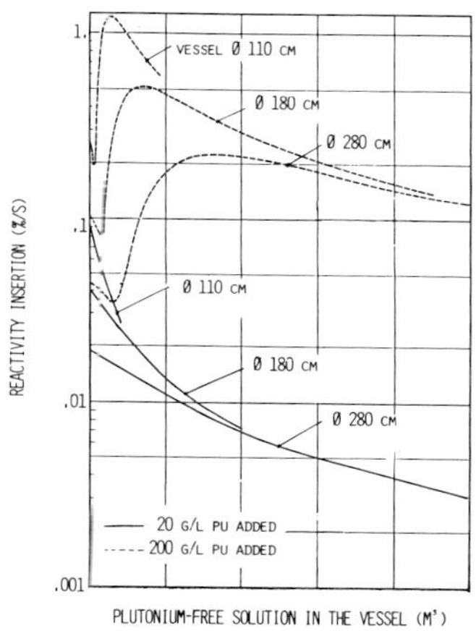

Fig. 2: Reactivity insertion versus time at an inadvertent transfer of plutonium nitrate solution in vessels of different sizes.

4 MODELLING OF CRITICALITY EXCURSIONS

4.1 MODEL DESCRIPTION

A quasi-steady state model was developed to describe the long-term behaviour of criticality /14/. Here, for a given temperature and a given concentration of fissile material a stationary neutron flux and a stationary flow rate of radiolytic gas or vapour are assumed. Hence the fast changes of the neutron flux and gas flow rate are neglected. Only the slow changes in temperature, concentration of fissile material and liquid volume due to the released nuclear power and the added liquid are described as a function of time. The reactivity is assumed to be zero at any time (Fig. 3).

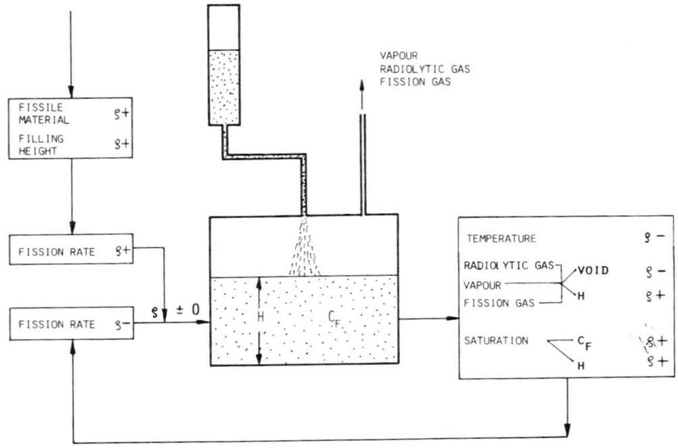

Figure 3: Overall scheme of the quasi-steady state modelling

With the above assumption the nuclear power can be calculated from the criticality equation. The dependence of the criticality equation on the concentration of fissile material, solution volume, void, temperature and nuclear power was modeled. There is a significant influence of the formation of void on the control of nuclear power. In a nonboiling liquid, the void is calculated from the radiolytic gas formation which is proportional to the nuclear power (G-values of H_2, O_2) and from the rising velocity of single bubbles, estimated from CRAC experiments /5/. When the boiling state is reached, the nuclear energy is exhausted by evaporation of water. Due to the volumetric heating of the solution, two aspects must be distinguished: partly suppressed and fully developed nucleate boiling. The later only occurs if the liquid is superheated by about 29° C. The minimum power Q_{min} for fully developed nucleate boiling can be calculated by means of heat transfer measurements of a volumetrically heated liquid performed by Fieg /15/ as a function of the vessel geometry:

$$Q_{min} = Fr \cdot \frac{R^2}{H^{0.2}} \qquad \begin{array}{l} Fr \sim 70 \text{ if } H/2R = 1 \\ FR \sim 35 \text{ if } H/2R = 0.25. \end{array}$$

When nucleate boiling begins, the nuclear power decreases considerably due to the increase in void. Therefore, the nuclear power is mostly in a range where a fully developed nucleate boiling cannot occur. Only in the case of a significant addition of fissile material after the solution has become critical and in connection with low filling heights H/D <<1 the nuclear power can reach values where fully developed nucleate boiling takes place.

The variation of power with time results from the heat-up of the solution, the variation of fissile material concentration and the variation of solution volume according to addition or evaporation of solution. The respective timedependent balance equations were solved numerically by using an explicit finite-difference method of first order. Thus, the model provides time-dependent results for nuclear power, energy, off-gas and vapour flow, void fraction, temperature, concentration of fissile material and other system-specific parameters.

The references /14/ and /16/ contain a more detailed description of the model.

4.2 VERIFICATION AT CRAC EXPERIMENTS

In order to verify the model it was applied to some experiments at the CRAC facility. Two experiments have been selected for discussion here (Table 1).

Table 1: **DATA OF CRAC EXPERIMENTS**

	CRAC 16	CRAC 18
Vessel Diameter	300 mm	300 mm
Uranium Concentration (93 % U-235)	82.3 g/l	79.6 g/l
Critical Liquid Level	42 cm	42 cm
Critical Volume	40 l	40 l
Total Volume Introduced	54 l	68.8 l
Flow Rate of the Liquid	443 l/h	102 l/h
Duration of Experiment	12 min	6 h

The CRAC 16 experiment /17/ was conducted in a vertical, cylindrical vessel of 300 mm diameter. At a height of 42 cm the solution of highly enriched uranyl nitrate became critical, causing the temperature to rise within about 5 min to the saturation temperature. The power was unsteady during the heat up (Fig. 4). When the solution started boiling the oscillations damped out. 12 min after becoming critical the excursion was terminated by draining the liquid. Due to the approximations of the model the oscillations could not be reproduced, but the integrated mean power agreed well with the observed increase of the total energy. The model predicts that approximately 10 % of the vapour contributed to the increase of void in the solution while 90 % evaporated from the surface.

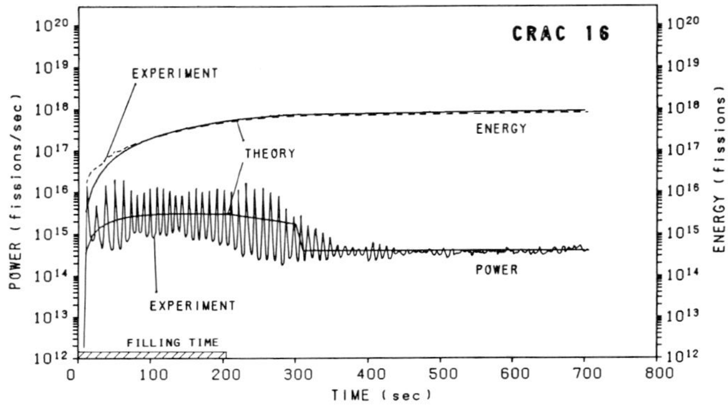

Figure 4: History of power and total energy released during the experiment CRAC 16 and comparison with the model

The CRAC 18 /18/ experiment is a typical example where the nuclear power was not reduced when boiling started (Fig. 5). Compared to the above-mentioned experiment, the difference lies in a lower flow rate but a higher quantity of liquid added. The boiling temperature was reached after 20 min. It can be derived from this model that only 5 to 10 % of the produced vapour contributed to the formation of void. During boiling, the nuclear power is similar to that in the CRAC 16 experiment because the power mainly depends on the vessel diameter only.

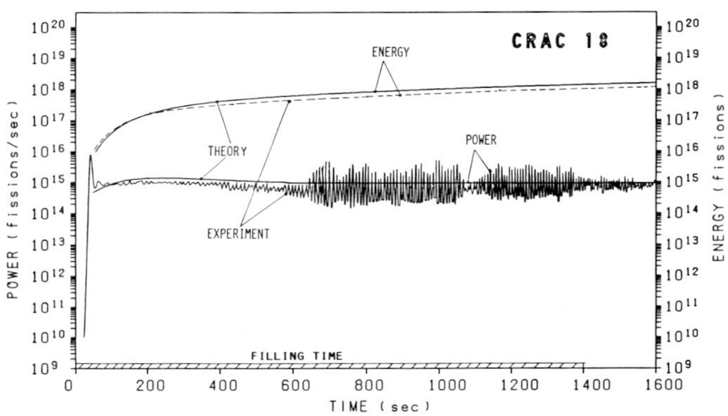

Figure 5: History of power and total energy released during the experiment CRAC 18 and comparison with the model

5 APPLICATION TO POSSIBLE CRITICALITY EXCURSIONS IN NUCLEAR REPROCESSING PLANTS

The possible criticality excursions in nuclear reprocessing plants caused by an inadvertent transfer of plutonium nitrate solution was parametrically analysed by means of the mathematical model. The following parameters were varied:

- vessel size (1, 5, 20 m^3 standard vessel)
- vessel of plutonium-free solution in the vessel before an inadvertent transfer
- volume of added plutonium nitrate solution after the solution becomes critical.

The concentration of the transferred plutionium solution is assumed to be 20 g/l with a feed rate of 1000 l/h. The inadvertent transfer is made into an empty vessel. The addition of liquid is terminated five minutes after the solution has become critical. Fig. 6 gives the fission rate and total fission number for three cases of inadvertent transfer of plutonium nitrate solution into vessel of different size. Any case of criticality is terminated by itself by evaporation of water. In the small vessel, the liquid must evaporate considerably until criticality is terminated. There are only minor differences in the total fission numbers. They range between 2.10^{19} and 5.10^{19} fissions.

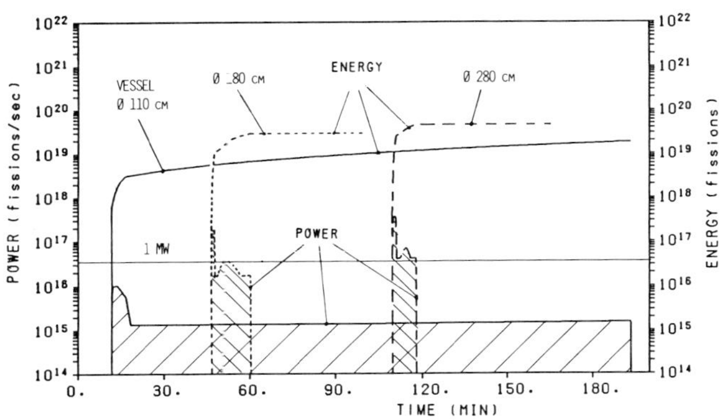

Figure 6: Calculated history of power and total energy released from possible criticality excursions in vessels of different size

The fission rate and the duration of criticality, however, depend on the vessel diameter. For the small vessel, the nuclear power reaches 50 kW during boiling, for the large vessel 1 MW. For the large vessel the criticality lasts several minutes, for the small vessel some 3 hours.

LITERATURE

/ 1/ J. Weber, W. Denk
Weiterentwicklung und Verifikation des Rechenprogramms FELIX zur Simulation von Kritikalitätsexkursionen.
Schriftreihe Reaktorsicherheit und Strahlenschutz, Bundesminister des Innern, BMI-1985-081, Juni 1985

/ 2/ E. Hörmann, J. Döhler
Sicherheitsanalyse von Prozeßschritten der Wiederaufarbeitung.
Fachband 2 zum Abschlußbericht des Projektes Sicherheitsstudien Entsorgung (PSE), Januar 1985

/ 3/ Nuclear Incident at the SL-1 Reactor on January 3, 1961 the National Reactor Testing Station.
U.S. Atomic Energy Commission, IDO-19302, January 1962

/ 4/ The Chernobyl Accident.
Nuclear News, Special Report
American Nuclear Society, June 1986

/ 5/ P. Lécôrché, R. L. Seale,
A Review of the experiments Performed to Determine the Radiological Consequences of a Criticality Accident.
U.C.C. Y-CDC-12, Oak Ridge, Tenn., 1973

/ 6/ W. R. Stratton
A Review of Criticality Accidents.
Los Alamos Scientific Laboratory, LA-3611, 1967

/ 7/ T. G. Hughes
Criticality Incident at Windscale.
Nuclear Engineering International, Vol. 17, February 1972

/ 8/ W. R. Casto
ICPP Criticality Event of October 17, 1978.
Nuclear Safety, Vol. 21, No. 5, 1980

/ 9/ W. L. Ginkel et al.
Nuclear Incident at the Idaho Chemical Processing Plant.
U.S. Atomic Energy Commission, IDO-10035, 1960

/10/ M. S. Dunenfeld, R. K. Stitt
Summary Review of the Kinetics Experiments on Water Boiler.
Atomics International, NAA-SR 7087, 1963

/11/ F. Barbry, J. C. Manaranche
Criticality Accident Experimental Studies with High-enriched Uranium Solutions.
Trans. Am. nucl. Soc. 41, 358-359, 1982

/12/ Wiederaufarbeitungsanlage Wackersdorf;
Sicherheitsbericht.
Deutsche Gesellschaft für Wiederaufarbeitung von Kernbrennstoffen mbH, Hannover 1983

/13/ W. Heinicke, Ch. Müller, W. Thomas, R. Warnemünde, W. Weber
Handbuch zur Kritikalität.
Gesellschaft für Reaktorsicherheit, 1979

/14/ T. Schulenberg, J. Döhler
Heating-up, Nucleation and Boiling of a Critical Solution of Fissile Material.
International Journal of Multiphase Flow, Vol. 12, No. 5, 1986

/15/ G. Fieg
Heat Transfer Measurements of Internally Heated Liquids in Cylindrical Convection Cells.
Eur. Appl. Res. Rep. 1(6), 1978

/16/ T. Schulenberg, J. Döhler
Modelling of Criticality Excursions in Aqueous Solutions of Fissile Materials.
ENC '86, Geneva, June 1986

/17/ F. Barbry, J. C. Bouly, D. Mangin, H. Revol
Expériences CRAC 15, 16, 17 Résultats.
SEEC No. 74, Commisariat a l'Energie Atomique, France, 1970

/18/ F. Barby et al.
Expériences CRAC 18, 19 Résultats.
SEEC No. 77, Commisariat à l'Energie Atomique, France, 1970

FELIX – a computer code for simulation of criticality excursions in liquid fissile solutions

B. Gmal, J. Weber

SUMMARY

Knowledge of characteristic parameters like evolved power fission yield during an accidental excursion is of essential importance to estimate possible radiological consequences and resulting safety hazards.

The computer code "FELIX" simulates excursion characteristics of aqueous critical assemblies: Starting out from given initial conditions the space-dependent neutron kinetic equations are solved in one-dimensional geometry. Power, fission yield, reactivity and temperature are calculated as a function of time. Reactivity-feedback includes density effects and radiolytic gas voids. Results from calculations are compared with CRAC-experiments.

B. Gmal, J. Weber, Gesellschaft für Reaktorsicherheit (GRS) mbH, D-8046 Garching, Federal Republic of Germany

1 THE IMPORTANCE OF CRITICALITY ACCIDENT ANALYSES IN THE LICENSING PROCEDURE OF NUCLEAR REPROCESSING PLANTS

The criticality safety of nuclear reprocessing plants is in general ensured by applying the double contingency principle. This leads to requirements that "process designs should, in general, incorporate sufficient factors of safety to require at least two unlikely, independent and concurrent changes in process conditions before a criticality accident is possible" /1/.

It is evident, that all measures are taken to avoid criticality accidents and the efforts are justified by success, since no criticality accident occurred in the F.R.G. up to now.

As a special feature of the German licensing procedure for fuel cycle facilities the potential radiological consequences of accidental nuclear criticality have to be evaluated. Furthermore it must be shown that the legal radiation limits at observation points outside the plant area will not be exceeded in the case of an accident.

A prerequisite for this evaluation is to establish conceivable accident scenarios. In a first step possibilities for a criticality violating the double contingency principle of safety are identified by screening the equipment and operation of the facility. Undetected accumulation of fissile material or incorrect transfer of fissile solution to unfavourable geometry normally are most important. In a second step relevant and credible scenarios causing the most severe consequences are derived from these possibilities and their possible consequences are investigated. Therefore calculational methods are necessary, which allow to determine the development of the identified relevant scenarios. Starting at defined initial conditions (type and concentration of fissile material, volume and shape of vessel etc.) the essential quantities of critical excursions must be calculated:

- time dependence of fission rate and power release
- characteristics of the first peak
- total number of fissions.

Using these data in the next step the radiological consequences (radiation fields of neutrons and gammas, release of fission products, formation of aerosols, pressure build-up) can be estimated.

2 EXPERIENCE ON CRITICALITY ACCIDENTS

At first, reliable data on critical excursions have been provided by the evaluation of real accidents. We know eight criticality accidents which happened in fuel reprocessing plants until now and all of them occurred after failure in handling or transfer of liquid fissile solutions /2/. Therefore we direct our attention on this type of fissile material systems.

Secondly experiments have been conducted to investigate the behaviour and development of critical excursions at different conditions. Of importance for fissile solution systems are the "Kinetics Experiments on Water Boilers, KEWB" /3/ carried out in the USA between 1958 and 1960, the French CRAC experiments /4/ (1968 to 1971) and the most recent experiments carried out since 1974 at the SILENE liquid fuel pulse mode reactor at Valduc, France /5/.

In a critical power excursion the main fraction of released energy (0,32 x 10^{-10}Ws per fission) occurs as heat (about 90%) and radiation. In very fast increasing excursion also mechanical shock waves were observed. The dominant hazard is caused by the emitted radiation, hard gamma radiation in the range of 1-10 MeV and neutrons. These quantities are determined by the time dependent evaluation of the fission rate. Besides the immediate radiation level the release of fission products (noble gases, iodine, aerosols) is responsible for possible radiological consequences.

Summarizing some essential results of these experiences we can state:

- Paramters influencing the development of a criticality excursion are material data (fissile material, chemical compositions, concentration), vessel geometry, addition rate of fissile material, initial power.

- An increase in temperature and occurence of gas microbubbles, e.g. by radiolytic dissociation of the solvent causes a decrease in reactivity.

- In supercritical systems with high excess reactivity and short rise time to the first maximum ($\lesssim$ 10 msec) peaks in pressure and acceleration up to 42 bar and 130 g have been measured.

- Radiolytic gas production was observed above an energy threshold of $1{,}5 \times 10^{12}$ fissions per cm^3.
- From the experimental data formulas have been derived to calculate the characteristic quantities of excursions using experimental parameters.

3 THE COMPUTER CODE FELIX

As experiments are very difficult to perform and only a restricted bandwidth of conceivable boundary conditions can be covered, computer simulation programs are needed, which - after validation by experiments - can deal with situations not accessible by experiments (e.g. Pu-solution as fissile material).The computer code FELIX (Fission Excursions in Liquid Solutions) has been developed to simulate the dynamic behaviour of criticality excursions in aqueous solutions of fissile material /6,7/. The main goal was to use as few adjustable parameters as possible. Instead the calculations are based on known initial conditions and relevant physical equations.

We use a microscopic model to describe the time dependent evalutation of the power release in a (super-) critical system. This leads to the wellknown problem of the time and space dependent neutron density represented by the Boltzmann equation. In this model the calculation is devided into two steps:

In a first step the static transport equation is solved in a one-dimensional approximation to give the neutron flux distribution averaged over volume and energy. This calculation is realized by using the ANISN transport code and the 16 group Hansen Roach cross section library, taking into account the actual material composition (fissile material, isotopic composition, concentration, acidity, impurities). Furthermore in this step the effective neutron multiplication factor k_{eff}, delayed neutron fractions β_{eff} and neutron life time Λ are calculated. These parameters will be used in the following dynamic calculation. The finite height of the fissile material zone is considered in a buckling correction.

In a second step the dynamic behaviour of the system is calculated using a set of coupled differential equations (a detailed representation is given in /6/):

- The reactor kinetics equations represent the time-evaluation of the neutron density which depends on the previous calculated neutron paramters β_{eff}, Λ and the reactivity R
- equation for heat conduction in the solution
- change in solution density due to temperature rise and appearence of microbubbles
- reactivity change.

The equations are coupled through the reactivity defined as

$$R = \frac{k_{eff} - 1}{k_{eff}}$$

which is the feed-back parameter in the model. Initial condition is an excess in reactivity ($k_{eff} > 1.0$), e.g. reached by filling a vessel with fissile material solution to a level above the critical height. This leads to an increasing fission rate. Rising temperature (and gas bubbles) yield a decrease of solution density and hence a negative reactivity change. Thus the time dependence of R is given by

$$\frac{\partial R}{\partial t} \sim \frac{1}{\rho}\frac{\partial \rho}{\partial t} + \frac{\partial R_{ex}}{\partial t}$$

ρ = density of the solution
R_{ex} = externally added reactivity

To describe the time behaviour of the gas bubbles in the fluid a simple exponential law with decay constant is used. The experimentally determined threshold value of $1{,}5 \times 10^{12}$ fission/cm^3, above which radiolytic gas production occured and the measured production rate of $1{,}5 \times 10^{-13}$ cm^3/fission are used in the calculation.

The equations are solved time-step by time-step and so the dynamic changes of the system can be simulated just by using common physical laws and constants (see flow diagram, Fig. 1). The only adjustable parameter used describes lifetime of gas bubbles in the solution. The

solution temperature as a function of time, thus allowing assessment of transient behaviour.

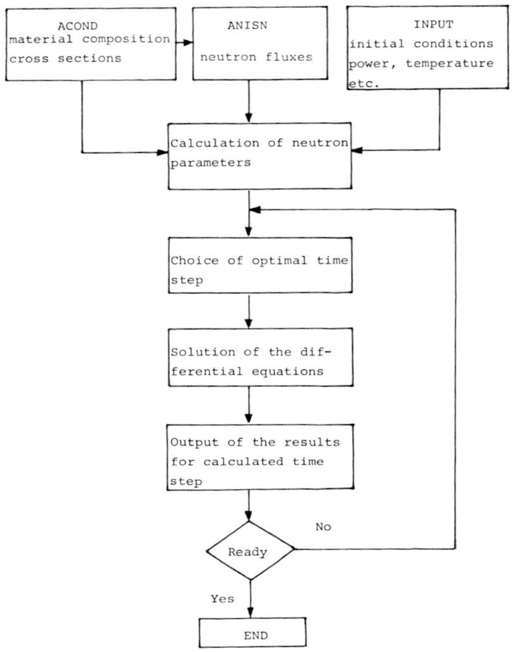

Fig. 1: Flow diagram of the calculation model FELIX

4 VALIDATION OF THE CODE FELIX

Up to now the program has been validated on seven CRAC-experiments. In all these cases a cylindrical vessel (diameter 300 mm) filled with uranyl nitrate (93% enriched uranium) was used. The U-concentrations varied from 50 g (U)/l to 200 g (U)/l, the reactivity addition rates from 1.4 x 10^{-3} \$/s to 0.79 \$/s, covering a wide range of reactivity insertion rates. Despite the existing uncertainties (e.g. initial power and acidity are often not known) power and fission yields are reproduced quite well by calculation with a tendency of overestimating the experimental results (table 1). In addition, the time dependence is reproduced satisfactorily too. Figure 2 shows an example of a low reactivity addition rate.

	time 1. peak		power 1. peak		energy 1. peak	
	CRAC	calculation	CRAC	calculation	CRAC	calculation
	s	s	fiss./s	fiss./s	fiss.	fiss.
Crac 01	232	303	$1,1 \cdot 10^{16}$	$1,4 \cdot 10^{16}$	$2,7 \cdot 10^{17}$	$7,4 \cdot 10^{17}$
Crac 03	427	469	$4,7 \cdot 10^{15}$	$7,6 \cdot 10^{15}$	$1,7 \cdot 10^{17}$	$5,6 \cdot 10^{17}$
Crac 04	197	237	$8,7 \cdot 10^{15}$	$1,3 \cdot 10^{16}$	$2,2 \cdot 10^{17}$	$6,8 \cdot 10^{17}$
Crac 05	21,6	16,2	$6,3 \cdot 10^{16}$	$1,4 \cdot 10^{17}$	$6.3 \cdot 10^{16}$	$6,6 \cdot 10^{16}$
Crac 06	22,8	15,9	$6,6 \cdot 10^{16}$	$1,7 \cdot 10^{17}$	$6,7 \cdot 10^{16}$	$7,2 \cdot 10^{16}$
Crac 07	3,9	1,4	$5,0 \cdot 10^{18}$	$1,5 \cdot 10^{18}$	$4,0 \cdot 10^{16}$	$3,4 \cdot 10^{16}$
Crac 13	7,5	7,9	$5,3 \cdot 10^{17}$	$5,6 \cdot 10^{17}$	$5,2 \cdot 10^{16}$	$5,2 \cdot 10^{16}$

Table 1: Modelling of CRAC-experiments by the computer code FELIX

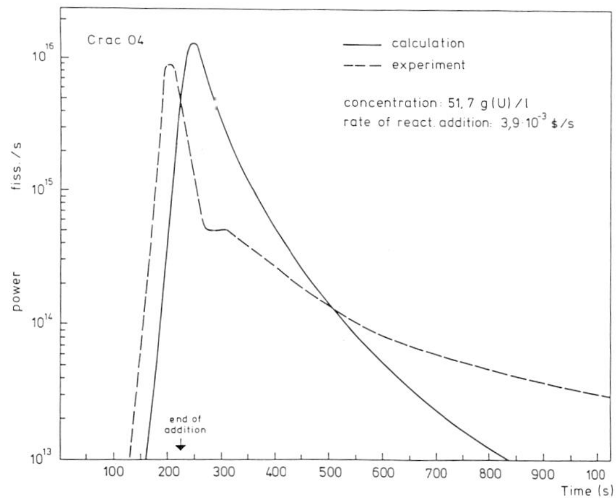

Fig. 2: Calculated and experimental values of fission rate for experiment CRAC 04

The different asymptotic behaviour of power decrease after the first maximum of experiment and calculation we attribute to the unspecified rate of heat loss of the system. Figure 3 shows a representative excursion with high reactivity insertion: The first maximum is followed by a series of power oscillations with decreasing amplitude. The periodic decrease in power is caused by the appearance of radiolytic gas bubbles. The period depends on the lifetime of the bubbles in the system. Lacking further information this parameter has been derived from the experiment. After approximately 150 s the oscillations are damped out and the power decreases slowly. Experiment and calculation show the same qualitative behaviour.

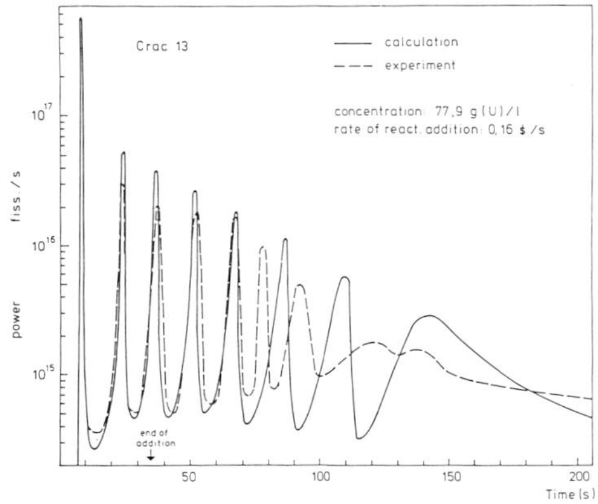

Fig. 3: Calculated and experimental values of fission rate for experiment CRAC 13

5 CONCLUSION

The simulation program FELIX shows good agreement with the calculated CRAC experiments ranging from "slow" to "fast" excursions without stressing arbitrary fit parameters. Starting with initial conditions of an excursion the characteristic quantities can be determined with satisfactory accuracy in the time range of some hundred seconds. This strenghtens the confidence that it can predict peak power and energy release during a hypothetical excursion even in cases where initial conditions are not backed by experiments.

LITERATURE

/1/ American National Standard, ANS-8.1, 1975

/2/ W. Thomas
Kritikalitätsunfälle, Übersicht 1945-1983
GRS-A-914, November 1983

/3/ M.S. Dunenfeld, R.K. Stitt
Summary Review of the Kinetics Experiments on Water Boilers,
Atomics International NAA-SR-7087, Feb. 1963

/4/ F. Barbry et al.
Recapitulation des Resultats Experimentaux, Section d'Etudes Expérimentales des Sureté Nucléaire et Criticité, Rapport Nr.116, 1973

/5/ F. Barbry
Fuel Solution Criticality Accident Studies with the SILENE-Reactor, International Seminar on Criticality Studies Programs and Needs, Dijon, September 1983

/6/ J. Weber, W. Denk
FELIX - ein Computerprogramm zur Simulation von Kritikalitätsexkursionen wässriger Spaltstofflösungen, BMI-1984-040, Juni 1984

/7/ J. Weber, W. Denk
Weiterentwicklung und Verifikation des Rechenprogramms FELIX, BMI-1985-081, 1985

Plutonium accumulation in MINKA experiments

J. Schön, M. Kluth, H.-J. Bleyl, H.-J. Schmidt

SUMMARY

The paper shows the influence of the acid concentration of the scrubbing solution and the impact of the fraction of dispersed phase on U-Pu accumulation in the S solumn.

If the dynamic equilibrium is attained, the experimental results can be simulated satisfactorily by the VISCO computer code.

Moreover, for the flowsheet characterized by a double disturbance (low acid concentration of the scrubbing solution and minor extractant flow) the experimentally determined behavior in terms of time of U and Pu accumulation is presented and teh differences in reaction heat during accumulations in the A and S columns are pointed out.

J. Schön, M. Kluth, Dr. H.-J. Bleyl, Kernforschungszentrum Karlsruhe, Institut f. Heisse Chemie, Postfach 36 40, D-7500 Karlsruhe 1, Federal Replublic of Germany
Dr. H.-J. Schmidt, Wiederaufarbeitungsanlage Karlsruhe Betriebsgesellschaft mbH, Postfach 2 20, D-7514 Eggenstein-Leopoldshafen 2, Federal Replublic of Germany

1 INTRODUCTION

At the Hot Chemistry Institute of the Karlsruhe Nuclear Research Center the MINKA (Mininatur-Pulskolonnen-Anlage) miniature pulsed column facility was set up in order to conduct experimental studies of PUREX flowsheet variants.

It is possible to follow experimentally in this facility the development with time of the temperature and concentration profiles over the column length. These studies can be performed under conditions according to specification as well as with the flowsheet conditions disturbed.

This paper reports on the conditions under which and the extent to which plutonium may accumulate in case of combined operation of the HA and HS columns.

The VISCO computer code can be verified using the data determined in these experiments. This allows reliable predictions to be made about the accumulation behavior of plutonium in the industrial-scale reprocessing plant.

2 DESCRIPTION OF THE FACILITY

2.1 COLUMNS

MINKA includes a complete extraction cycle consisting of four pulsed columns which can be connected to make up a codecontamination cycle, a separation cycle or a cleaning cycle.

The active height of the columns is 3.5 m. They are equipped with 0.5 m spaced sampling points for the organic, aqueous and mixed phases as well as with a temperature measuring point.

The A and S columns have been designed for continuous operation with the organic phase, the B and C columns for continuous operation with the aqueous phase. Further data are given in table 1.

The feed streams of the S and C columns and the two columns themselves can be heated up to 60° C at the maximum.

Table 1: Data of the MINKA columns

Column	A	S	B	C
function	extraction	scrub	partitioning	strip
continuous phase	organic	organic	aqueous	aqueous
height [cm]	350	350	350	350
diameter [cm]	2.6	2.6	3.5	3.5
heating [° C]	no	25-60	no	25-60
Range of pulsation				
frequency [H_2]	0.25-2.5	for all columns		
amplitude [cm]	0.5·1.5	0.5·1.5	1-2.5	1-2.5
Sieveplate				
% free area	23	23	23	23
space [cm]	3	3	5	5
holes [cm]	0.15	0.15	0.3 (Nozzle)	0.32

2.2 PULSATION

All four columns are provided with a pulsator hydraulically driven by a reciprocating pump. The pulsator is connected with the radioactive inventory of the pulsed column via a remote head membrane part. This system allows to set a frequency range of 15-150 l/min. and an amplitude range up to 25 mm at the maximum. The pulsation and the dispersed phase fraction are monitored continuously by pressure cells installed at the foot of the column.

2.3 PHASE INTERFACE

The phase interface in the continuously operated A and S columns for the organic phase occurs at the column foot. Interface control in the A column is effected with the help of a probe; by combination of conductivity and capacitance it delivers a nominal flow which is proportional to the level of the interface. This system is used also for interface control of the continuously operating C column with the aqueous phase. In the B and S columns an ultrasonic probe is used to detect the interface.

2.4 STREAMS

The process streams are monitored by instruments measuring the mass flow and volume flow. In order to be able to observe the required metering accuracy of < 1 %, so-called remote head membrane pumps have been installed in the active zone. The cold feed streams from the makeup are metered in via reciprocating pumps.

2.5 IN-LINE INSTRUMENTS

The plutonium content in the raffinate streams of the A and C columns is measured and recorded continuously by alpha monitors /1/. In the aqueous raffinate stream of the S column (SR) which is recycled into the A column the uranium and plutonium contents are measured and recorded every 10 minutes by a laser light pipe photometer /2/.

2.6 ADDITIONAL INFORMATION

The organic solvent is recycled; scrubbing is performed in two two-stage Holleymott mixer-settlers.

The valency of plutonium is set to conform to process conditions using the electrochemical apparatuses ELKE, ROXI and REX /3/.

Prior to being concentrated by evaporation the aqueous product streams are freed from organic substances entrained in sorption resin beds /4/.

All apparatuses and vessels have been installed in a glovebox of 30 m^3 volume, 6 m length and 5 m height.

3 EXPERIMENTAL

The extraction behavior of uranium, plutonium and nitric acid was studied in a codecontamination cycle. The flowsheet is shown in table 2.

The interconnection of the combined columns and the systems required in these experiments are visible from fig. 1.

The data of pulsed column flooding can be taken from table 3.
In the course of the experiments the consequences on the U, Pu and HNO_3 extraction behavior of the disturbances described below have been investigated:

Table 2: Conditions of the Codecontamination cycle

Flowsheet					
flow name	flow relative	composition			
		M HNO_3	g U/l	g Pu/l	TBP
AF	1.0	2.8-3.0	245-260	1.5-2.1	-
AX	2.8	0.4	< 0.006	< 10^{-3}	30
SS	0.4	0.78;0.4;0.2	-	-	-
CX	5.7	0.1 (40° C)	-	-	-

Columns	Frequency [HZ]	amplitude [cm]	through-put [l/h]
A	1.2	1.5	10-12
S	1.0-2.0	1.5	6-8
C	1.0 (40° C)	2.5	15-20

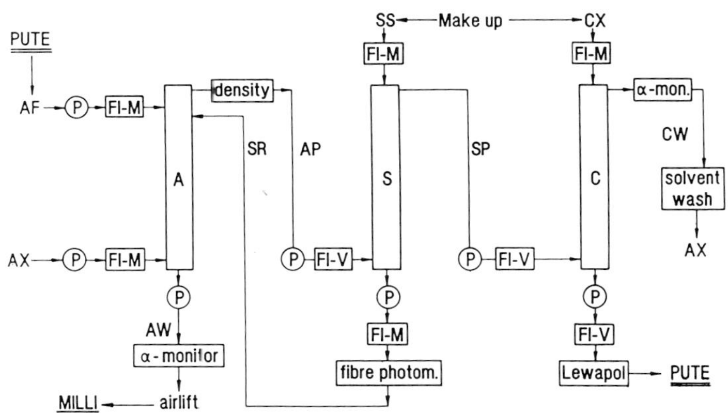

Fig. 1: Process flow diagram of the MINKA test facility

1. Disturbance:

Lowering the acid concentration in the scrubber column from 0.78 to 0.4 and 0.2 M HNO_3.

Table 3: Flooding points of the MINKA columns

Flowsheet	AF	$3\ M\ HNO_3$ 245 g U/l		
	AX	30 % TBP / Alkane		
	SS	$0.8\ M\ HNO_3$		
	CX	$0.1\ M\ HNO_3$		
Column		A	S	C
Flowratio $^O/_A$		2.5	7.0	0.46
amplitude cm		1.5	1.5	2.5
Flooding points [l/h] at Frequency Hz				
	0.83	-	-	28.3
	1.0	17.9	-	27.4
	1.17	17.5	-	-
	1.33	12.9	16.8	20.2
	1.5	-	14.9	-

2. Disturbance:

After the disturbance had been set again the fraction of aqueous dispersed phase (holdup) was increased in the scrubber column.

3. Disturbance:

After the acid concentration in the scrubbing solution had been lowered to $0.2\ M\ HNO_3$ another disturbance was superimposed to this disturbance, which was achieved by lowering the HAX flow rate from 100 to 95 % of the value of the nominal flowsheet, and the development with time of U-Pu accumulation was investigated in detail.

Backextraction of uranium and plutonium was performed in the C column which had not been included in the accident related studies. Initially, losses of product in the CW stream were observed of about 1 g U/l and 40-50 mg Pu/l. After the column temperature had been raised to 40° C and hydroxylammonium nitrate added as reductant to the $0.1\ M\ HNO_3$ CX solution, the losses decreased to < 6 mg U/l and to < 0.5 mg Pu/l.

4 RESULTS

1. Disturbance: Lowering the acid concentration in the scrubbing solution

During the first experiments the acid concentration of the scrubbing solution SS was lowered from 0.78 M HNO_3 (nominal flowsheet) to 0.4 M and finally to 0.2 M while the other parameters specified in the flowsheet (table 2) were kept constant.

After the equilibrium had been attained, the concentration profiles of plutonium, uranium and nitric acid were determined analytically. The maxium concentrations were measured in the aqueous phase of the S column.

Figure 2 is a plot of the uranium and plutonium concentrations in the aqueous phases along the column heights of the A and S columns for 0.78, 0.4 and 0.2 M HNO_3 in the scrubbing solution. The figure makes clear that uranium and plutonium accumulate only in the aqueous phase of the S column and that this accumulation decreases with the acid concentration in the scrubbing solution getting lower.

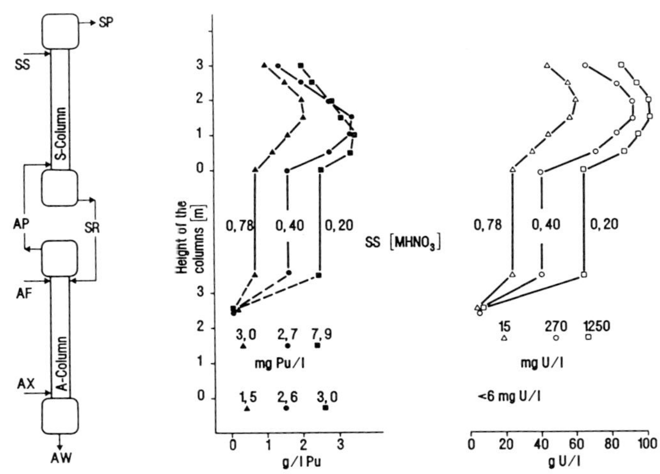

Fig. 2: Concentration profiles of U and Pu in the aqueous phase of the A and S columns at varying HNO_3 molarity

In table 4 the measured maximum concentrations for accumulation are summarized and related to the concentration of the feed solution. The table also shows the amounts of uranium and plutonium recycled with the raffinate stream SR and compares them with the amount supplied together with the feed solution.

Table 4: Accumulation of Pu and U with Decreasing Scrub Acid

nitric acid in scrub solution M HNO_3	0.2	0.4	0.78
Pu-accumulation			
max. g/l	3.2	3.2	1.9
% of feed conc.	145	150	92
U-accumulation			
max. g/l	99	90	57
% of feed conc.	34	36	23
mass recycling by raffinate SR % of feed input			
Pu	57	32	13
U	11	7	4

2. Disturbance: Variation of the holdup of the S column with the scrubbing solution reduced to 0.2 M HNO_3

In these experiments the influence exerted by the column hydraulics on U-Pu accumulation was studied.

The flowsheet conditions entered in table 2 were maintained; only the acid concentration of the scrubbing solution was reduced to 0.2 M HNO_3.

By stepwise increase in pulsation in the S column from 60/min to 120/min the holdup of the S column was increased from 2 % to 9.8 %.

The influence exerted by the rising holdup on Pu and U accumulation is visible from fig. 3: with increasing holdup rises of the maximum plutonium concentration and of the maximum uranium concentration are achieved in the aqueous phase of the S column. Moreover, the Pu maximum is found to become displaced towards the foot of the column. This behavior of plutonium is even more evident from fig. 4 which

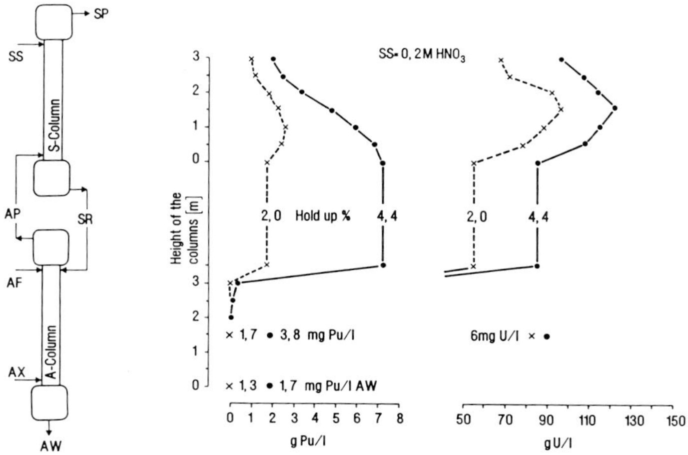

Fig. 3: The influence of hold-up on the concentration profiles of U and Pu in the A and S columns

shows only the plutonium concentration profiles of the aqueous phases of the S column.

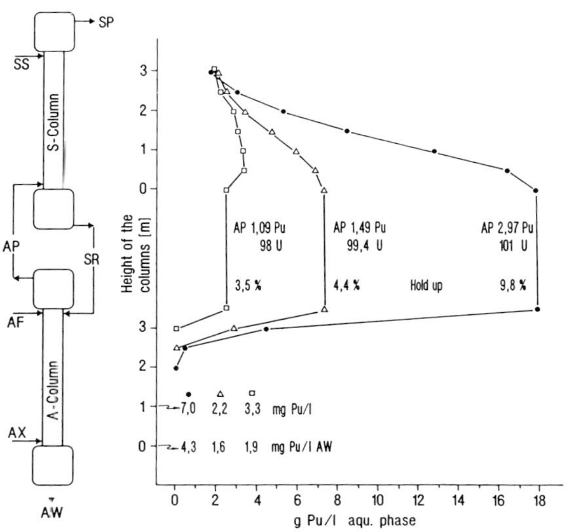

Fig. 4: The influence of hold-up on the concentration profile of Pu in the A and S columns

In table 5 the measured maximum concentrations of uranium and plutonium are compared with the feed concentrations, and the recycled amounts of uranium and plutonium, respectively, in the raffinate are given in proportion to the amounts of uranium and plutonium supplied to the feed.

Table 5: Accumulation as Function of Hold-up

% Hold-up S-column	2.0	3.5	4.4	9.8
Pu-accumulation				
max. g/l	2.5	3.2	7.2	17.8
% of feed conc.	175	145	820	1560
U-accumulation				
max. g/l	95	99	122	143
% of feed conc.	39	34	50	58
mass recycling by raffinate SR				
Pu	53	58	282	697
U	9.1	10.8	14.4	14.8

5 COMPARISON OF VISCO BASED COMPUTATION WITH EXPERIMENTAL DATA

The experimental data obtained of uranium and plutonium accumulations and their dependence on the acid content of the scrubbing solution and on the holdup fraction in the S column can be simulated well by the VISCO computer code, which is demonstrated in figure 5.

The figure shows the U, Pu and HNO_3 profiles for the experiment involving 0.2 M HNO_3 in SS and 3.5 % holdup. The U and Pu concentrations, respectively, expressed in g/l, and the HNO_3 concentration, expressed in mole/l, have been plotted versus the height of the combined A+S columns. This height is calculated from the active length of the A column, the height of the A decanter and the bottom decanter of the S column as well as the active length of the S column. In the combined columns the feed and raffinate streams are supplied at 3.5 m level. AX is supplied at 0 m level and the scrub enters at 7.8 m level. In figure 5 the measured values have been entered together with their symbols (= aqueous phase, o = organic phase) and the plots reflect the VISCO based computation.

3. Disturbance: double disturbance by simultaneous lowering of the HNO_3 concentration in the scrubbing solution to 0.2 M and reduction of

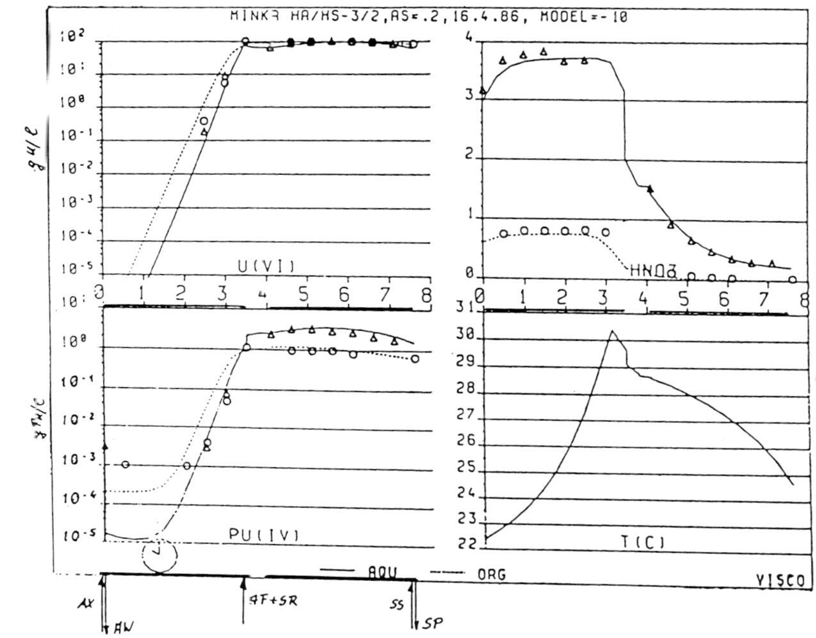

Fig. 5: Simulation of experimental concentration profiles of U, Pu, and HNO_3 in the S column by the VISCO computer code

the extractant flow to 95 % of the value specified in the nominal flowsheet (table 2).

At the end of 62 hours the experiment had largely attained equilibrium condition. The uranium and plutonium concentration profiles for the flowsheets have been entered in fig. 6 with 100 % and 95 % AX flows, respectively. The rise in the uranium and plutonium accumulation by reduction of the AX stream is clearly visible. However, under these conditions the accumulation is still limited to the S column. The recycled uranium and plutonium masses alone do not give rise to a disturbance in the A column.

The development with time of uranium and plutonium accumulations in the S column is visible from fig. 7 and from fig. 8. For reasons of transparency the uranium and plutonium concentrations, respectively, have been entered only for the zone of interest in the S column: 0 m level = SR discharge, 1 m level = S_3 and 1.5 m level = S_4. The AP values indicate the concentration of the organic feed at 0 m level and the SP values apply to the organic discharge of the S column at 3.5 m level.

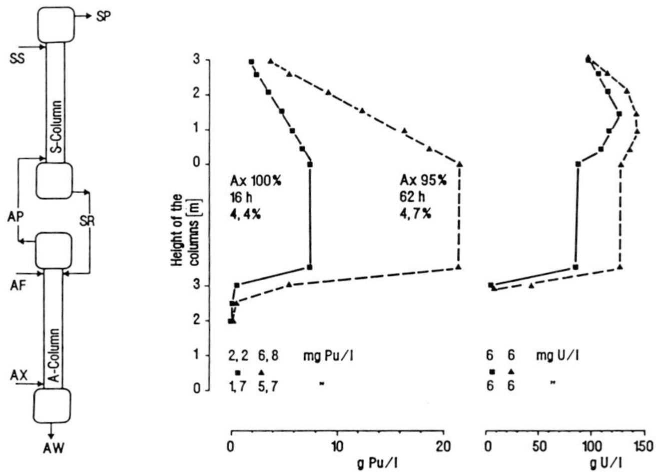

Fig. 6: Dependence of the U and Pu concentration profiles in the S column on the AX flow

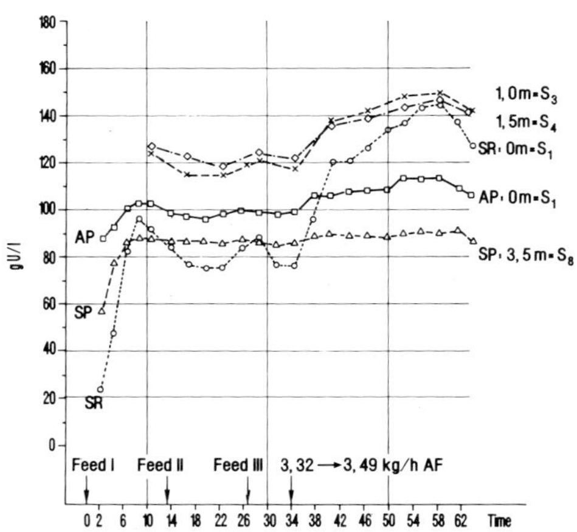

Fig. 7: Transient behavior of the U concentration in the S column

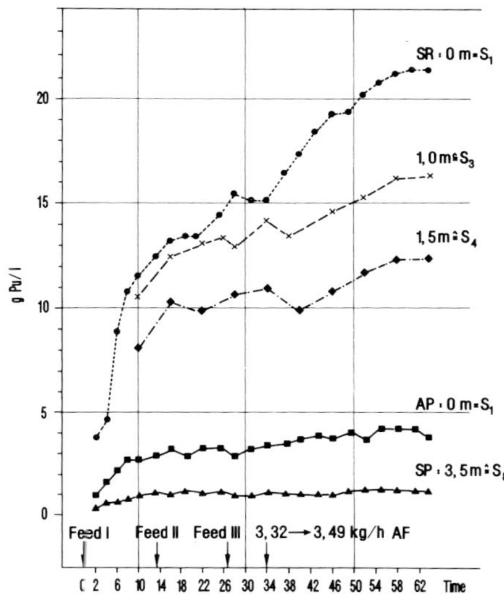

Fig. 8: Transient behavior of the Pu concentration in the S column

The discontinuous rise in U and Pu concentrations, respectively, is caused by the feed replacement required by the facility and by an increase in the feed flow, respectively.

The increase in feed flow became necessary in order to largely compensate differences in the facilities and in the experiments regarding the preparation of feeds.

This increase in feed flow and intensified uranium and plutonium recycling in the raffinate of the S column have led to a U-Pu accumulation in the A column which is limited in terms of time.

Figure 9 shows the development with time of the temperature and U concentration in the aqueous phase of the A column during accumulation. For this representation we have chosen the measuring points A6 = 1 m and A7 = 0.5 m below the feed supply which have been well recorded analytically. The U concentration of the recycled raffinate stream of the S column (SR) and of the AP stream have been entered in addition. The SR stream is fed together with the feed at 3.5 m level of the column. The organic AP stream leaves the A column above the point of feed supply via the TOP decanter. It is visible from the graphical representation that uranium and plutonium accumulated at the measuring point A7 have not been fully removed after 24 hours.

Plutonium accumulation paralleled uranium accumulation at the measuring points A6 and A7.

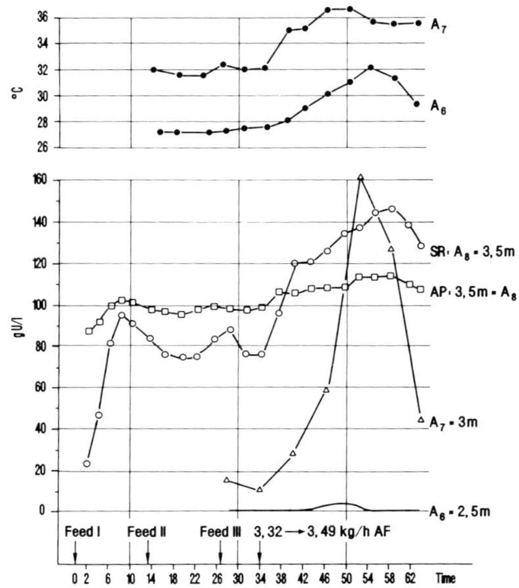

Fig. 9: Transient behavior of the U concentration and the temperature in the A column

The plutonium content of the aqueous phase A_7 increased from 1.8 g/l to a maximum of 10 g/l and was still 5 g plutonium/l at the end of the experiment.

At the point of measurement A_6 = 1 m below the AF+SR feed the concentrations during accumulation increased from 0.17 g Pu/l and 0.32 g U/l, respectively, to a maximum of 3 g Pu/l and 10 g U/l. At the end of the experiment the values were 0.71 g Pu/l and 1.4 g U/l.

The accumulation did not exert a noticeable influence on the measuring point A_5 = 1.5 m below the AF+SR feed. The concentrations of the aqueous phase here attained < 0.12 g U/l and < 0.1 g Pu/l.

It appears from a comparison of fig. 7 with fig. 9 that for uranium accumulating in the A and S columns approximately the same concentrations are measured in the aqueous phase. But only in the A column this accumulation is linked to a clear reaction heat which is due to the uranium extraction performed simultaneously. In the MINKA

experiments this led in the A column to a temperature rise from 4° to 5° C.

The development with time of temperature at the measuring points A_6 and A_7 has been plotted in the top part of fig. 9 versus the respective uranium concentrations.

Unlike in the A column, the temperature in the S column was rather uniform at 24° ± 1° C in the experiments.

ACKNOWLEDGEMENT

The authors would like to thank Dr. H. Schmieder for the discussion of the experimental results, Dr. G. Petrich for having performed the VISCO based computations, and Mr. U. Galla for his assistance in the combined operation of PUTE-MINKA. They also thank the MINKA and PUTE teams of operators for the excellent work done in carrying out the experiments. Thanks are due to Dr. L. Stieglitz and Dr. D. Ertel and coworkers for the analytical work done in support of the experiments.

LITERATURE

/1/ P. Groll et. al., KfK 3231 (1982)
/2/ J. Römer et. al., KfK 3844 (1984)
/3/ H. Schmieder, H. Goldacker, ISEC 86, Vol. I p. 117
/4/ W. Ochsenfeld, J. Schön, ISEC 83, p. 92

Pu-accumulations in the HA-extractor and a new flowsheet to avoid them

G.Petrich, H. Schmieder

SUMMARY

The conditions for accumulation of Pu were thoroughly studied with the aid of our Purex process computer model VISCO and a flowsheet was developed which reliably avoids the possibility of Pu accumulation and therefore facilitates any criticality provisions. The flowsheet was successfully operated in both mixer settlers and pulsed columns. Our experiments confirmed the prediction of a dramatically decreased Pu/U ratio in the aqueous raffinate HAW as compared to the traditional process conditions. The new flowsheet has the additional potential to increase fission product separation in the first cycle; first results are presented in this paper.

Dr. G. Petrich, Dr. H. Schmieder, Institut für Heisse Chemie,
Kernforschungszentrum Karlsruhe, Postfach 36 40, D-7500 Karlsruhe 1,
Federal Republic of Germany

The effect of Pu accumulations in the Purex first cycle high active extraction and scrub extractors HA and HS was first described in detail by Rozen [1] and Ochsenfeld et al [2]. A variety of process conditions may cause such accumulations: reduced solvent flow, reduced TBP concentration, increased high active feed concentrations or flow, low acidity etc. The Pu accumulations in a counter-current extractor occur by loading the solvent close to its theoretical limit. An insufficient solvent supply creates a "bottleneck" for organic Pu transport under usual flowsheet conditions and Pu is forced to the aqueous phase. Close to the raffinate loading is low and Pu is extracted again. This recycling allows Pu to accumulate. The smaller the solvent deficit, the slower the build-up of the Pu peak, but the larger its maximum concentration. A large solvent deficit causes a rapid propagation of the peak towards the aqueous raffinate outlet. The peak deflates rather rapidly when its tail reaches the raffinate. As a consequence the maximum peak concentrations increase with increasing extractor length.

Recently Schön et al. [3] observed peak concentrations of up to 40 g Pu/l (LWR feed composition, conventional flowsheet) in the miniature pulsed column facility MINKA (column length 350 cm). Theoretical calculations using the computer code VISCO [4] predict possible transient peak concentrations for technical size columns in excess of 80 g Pu/l under the most unfavorable conditions. Even though the build-up time of several days will hardly pass unnoticed a potential criticality problem exists and requires additional design restrictions or at least highly reliable safety instrumentation.

Standard Purex flowsheets for the HA/HS extractors favor the possibility of Pu accumulations: U(VI) has the highest extractability and expels all other species from the solvent at high loadings. Extensive studies using VISCO in order to minimize Pu accumulations predicted a possibility to completely avoid this risk. The overall HA extractor temperature or at least the temperature at the location of the "bottleneck" has to follow:

$$T \geq 401+(0.06676U-0.3367P-327.4H)\cdot \exp[0.00008179H\cdot(U+P)]\cdot H^{-0.9593} \quad \text{(equ.1)}$$

T is in °C, U and P are the aqueous U and Pu feed concentrations resp. in g/l, H is the HNO_3 feed concentration (M/l). This minimum extractor temperature does not depend on TBP concentration. It reflects the temperature and concentration conditions for which the distribution coefficient of Pu(IV) becomes larger than the distribution coefficient of U(VI). If the condition in equ.1 is met, Pu(IV) forms the strongest TBP-complex of all species involved. It therefore expels U(VI) from the organic phase and Pu cannot be recycled any more within the extractor: Pu accumulations in the HA extractor are reliably avoided. There may be uranium accumulations instead but even for FBR feed composition they will amount to only a few g/l and will be of no practical significance.

Equ.1 tells us that the critical temperature T_c (defined by equality of both sides) will be lower the larger the feed concentrations are. This is obvious from figs. 1 and 2 where the distribution

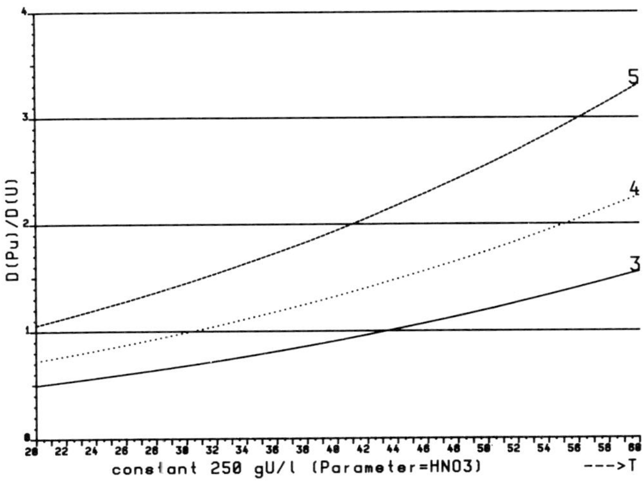

Fig. 1: Dependence of the ratio of the distribution coefficients D_{Pu}/D_U on temperature at constant U (250 g/l) and varying HNO_3 concentration (3,4 and 5 M)

coefficient ratios D_{Pu}/D_U are plotted as a function of temperature for 3, 4 and 5M HNO_3 at constant $[U(VI)]_{aq}$ = 250g/l (fig.1) and for 200, 250 and 300 gU/l at constant 4M HNO_3 (fig.2). Pu accumulations are avoided at D_{Pu}/D_U 1. A plot of D_{Pu}/D_U vs. total aqueous nitrate concentration at different temperatures as derived from the SEPHIS code [5] is published in [6] and exhibits $D_{Pu}/D_U>1$ at $[NO_3^-]>3.5M$ and T=55°C; this finding has not further been exploited to our knowledge.

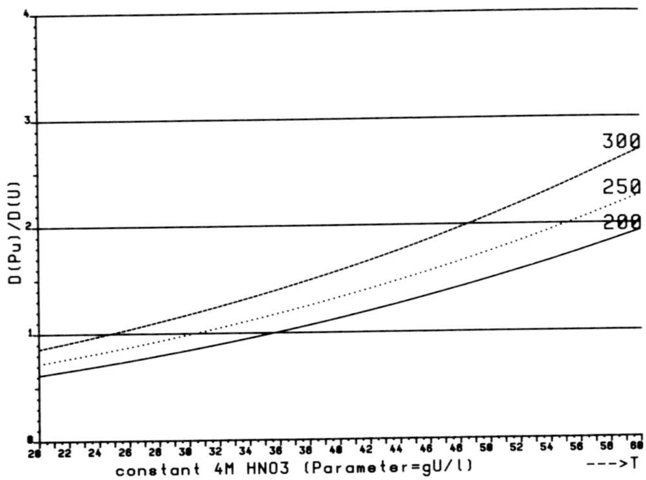

Fig. 2: Same as fig. 1 for constant HNO_3 (4 M) and varying U concentration (200,250 and 300 g/l)

Assuming solvent loading to the limit of solvent capacity, the "bottleneck" for a very small increase of solvent loading would be located at the stage of maximum uranium concentration. The HA uranium concentrations can nowhere exceed the feed concentration in this case, while minimum nitric acid concentration cannot be smaller than its feed value. This is the reason that only feed concentrations and no local concentrations enter equ.1. For composite extractors such as HA and HS the effective aqueous HA feed concentrations (e.g. HAF blended with HS-reflux) have to be inserted in equ.1.

Fig.3 Shows the U and Pu concentration profiles in g/l for a small lab scale 8-stage mixer settler operated at 50°C. HAX (30%TBP, 258ml/h) enters at stage 1, HAF (150ml/h, 211 gU/l, 2.35 gPu/l, 5.26M HNO_3) enters at stage 8. Solid curves are aqueous calculations, dashed curves are organic calculations, circles are organic measurements, triangles aqueous measurements.

Fig.4 shows measured and calculated concentration profiles for an operating temperature of 18-20°C with all other conditions kept identical to fig.3. While the feed composition U/Pu is 90, raffinate U/Pu is 2300(!) in the 50°C experiment shown in fig.3, and 60 in the low temperature test of fig.4. A further test run at 20°C and 3M HNO_3

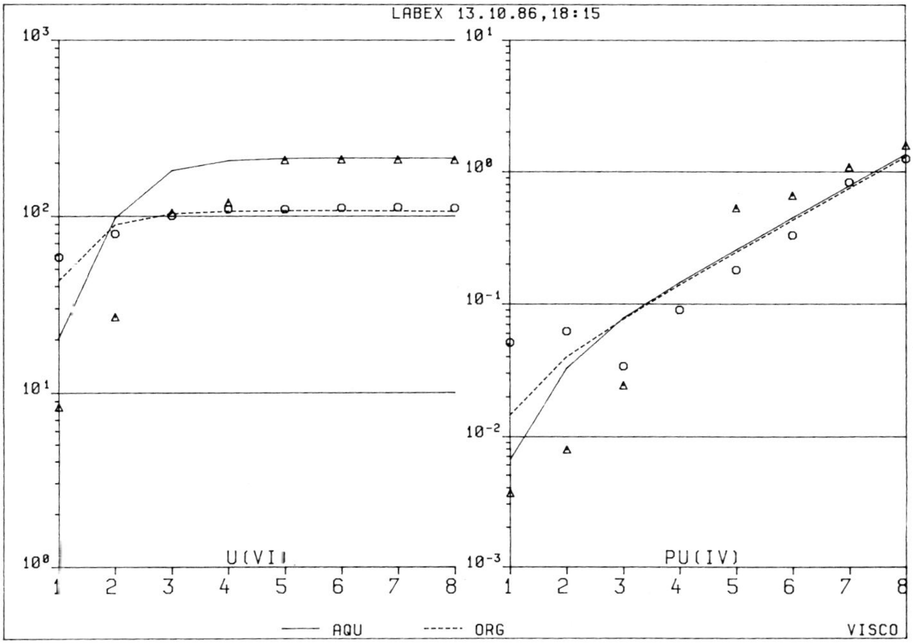

Fig. 3: U and Pu concentration profiles (g/l) in the 8 stages of a mixer settler at 50° C circles represent experimental data from the organic, triangles from the aqueous phase; solid and dashed lines represent calculations for the aqueous and organic phase, respectively

(T_C=45°C) increased Pu to U/Pu=15 in the raffinate and showed the expected Pu accumulation of 7g/l in stage 4. The 18-20°C experiment shown in fig.4 was operated almost at T_C=20.4°C. U and Pu show a very similar behavior and there is even a slight indication of a Pu peak. A further experiment in the 350cm A column of MINKA was run at a nominal 20°C using the feed composition of figs.3 and 4. But in this column the heat of extraction resulted in a temperature of 37°C at the location of the "bottleneck" which is in favor of the new process: no Pu accumulation was observed and the raffinate ratio was U/Pu 130.

These results demonstrate that not only Pu accumulations are reliably avoided by the new process but that in the case of a maloperation Pu is even depleted in the raffinate stream (it is by no means necessary to have U losses as large as in the tests of figs.3 and 4 to avoid Pu

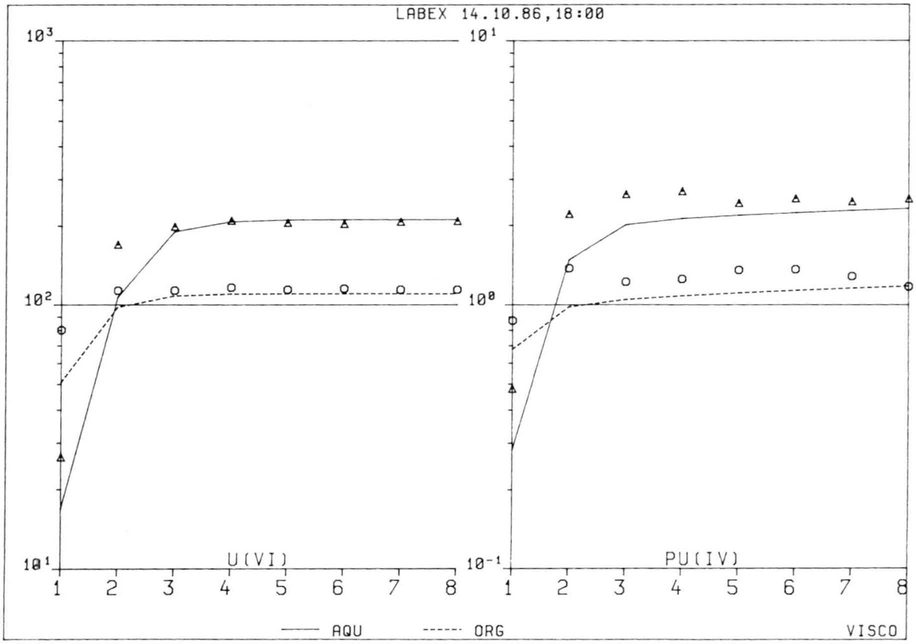

Fig. 4: Same as fig. 3, except that the temperature is 18-20° C

peaks). This allows to exploit the advantages of an ultra high loading process reducing the risk of Pu losses. As demonstrated in the ORNL work [7] the position of the U extraction flank can be kept constant by automatic control using suitable in-line sensors (temperature, light absorption etc.). Instead of keeping the flank position close to the feed point [7] it can be moved close to the raffinate, dependent on how large U losses can be tolerated. The reward is a highly enhanced fission product decontamination.

Fig.5 shows measured and calculated U and Pu profiles for an 8-stage HA extractor (stages 1-8) and an 8-stage HS extractor (stages 9-16). HA is operated at 50°C. HAX (29%TBP, 327 ml/h) enters at stage 1, HAF (145 ml/h, 230 gU/l, 2.87 gPu/l, 120 mgNp/l, 1 gZr/l, 1 gNb/l, 320 mgRu/l, 5.05M HNO_3) at stage 8, HAS (65 ml/h, 3M HNO_3) at stage 16. No Pu accumulation is observed in HA where the new process conditions apply (T_C=33°C). HS is operated at 25°C, while T_C for the HS extractor is 59°C and Pu accumulates indeed.

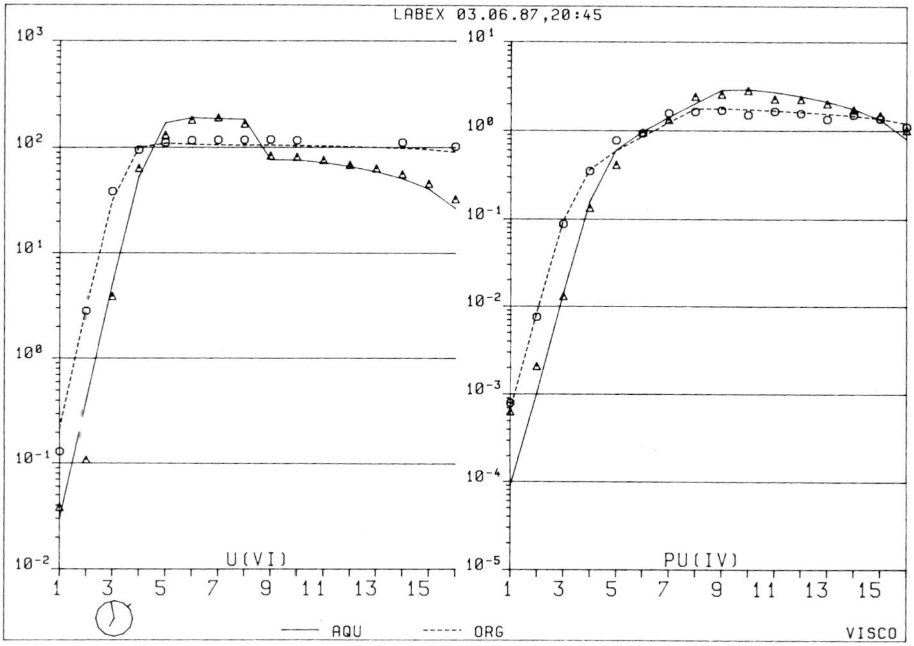

Fig. 5: Measured and calculated U and Pu concentration profiles for the 8 stages of an HA (1-8) and an HS extractor (9-16); for an explanation of the symbols refer to the legend of fig. 3

Fig.6 shows the measured concentration profiles of Zr and Nb. Ordinates are % of the feed concentration. Both species form weaker TBP complexes than U(VI) and therefore accumulate in the lower stages of HA where solvent loading decreases. The mechanism for this build-up is the same as for Pu(IV) in a conventional flowsheet as described above. Excellent decontamination is achieved since the tails of the accumulations do not extend to the organic product. Ru does not even accumulate and is found 100% in the raffinate as was to be expected.

In the example of figs.5 to 7 Np(VI) and Np(V) are reduced in HS predominantly to Np(IV) by operating the HS scrubber with electro-reduction at a current density of about 6 mA/cm . Under these conditions Pu(IV) and U(VI) are also reduced in small quantities but they are rapidly reoxidized in the HA extractor as can be seen from the small Pu losses to the aqueous raffinate. Fig.7 shows the measured concentration profiles along with Np calculated as Np(IV) (Fig. 7a)

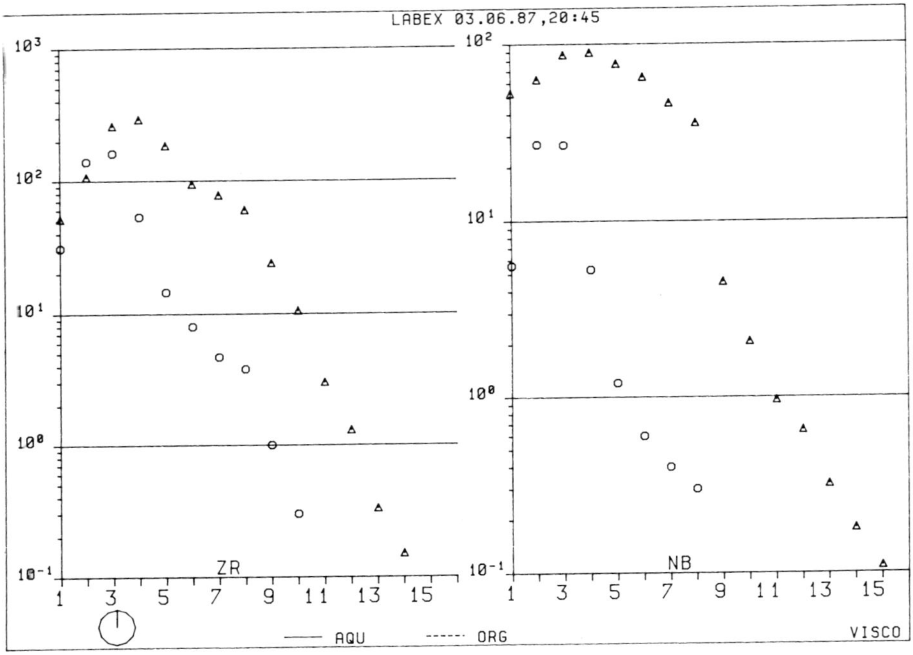

Fig. 6: Measured concentration profiles of Zr and Nb in the same mixer settler battery; for experimental conditions refer to the text

and Np(VI) (Fig. 7b); Np redox reactions are not yet implemented in VISCO. From the Np slope in the HS part (stages 9-16) of fig.7 it is obvious that little Np(VI) is present in the scrubber, whereas in HA Np is predominantly hexavalent. Np(IV) entering HA reoxidizes, the tailing in stages 1 and 2 is attributed to the almost inextractable Np(V) leaving with the raffinate. The Np accumulation is caused mostly by Np(VI) but due to electroreduction it cannot extend into the organic product: Np is quantitatively removed from the product.

Tc behavior was also studied in the new process. High temperature and high acidity are opposed to a good separation in the presence of Zr [8] and Tc was almost completely found in the product. Electroreduction did not reduce major amounts of Tc(VII) in the scrubber. Without Zr about 80% Tc was removed. Further experiments to reduce the Tc extraction factor at a lower organic to aqueous flow ratio are in preparation.

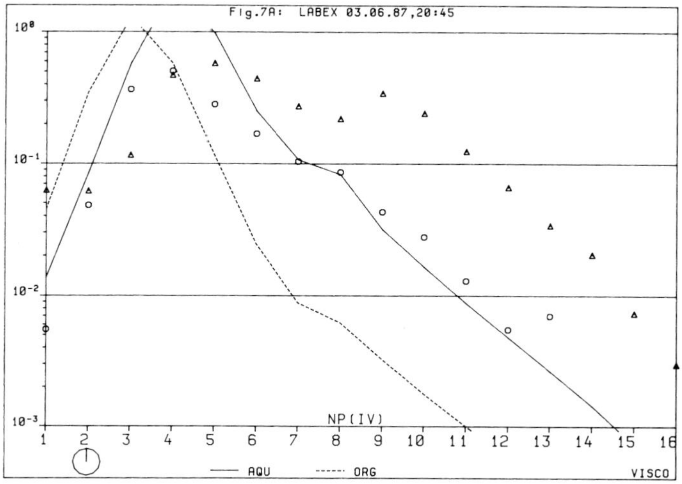

Fig. 7a: Experimental Np concentration profiles in the organic (circles) and aqueous phase (triangles) of an HA and HS mixer settler (8 stages each) and calculated profiles for Np(IV).

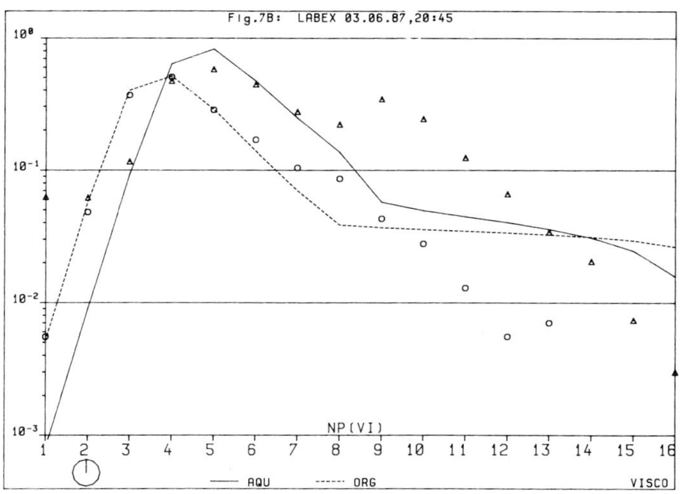

Fig. 7b: Same as fig. 7a, but Np distribution calculated as Np(VI)

Increased solvent degradation was expected for the new process [9]. The measured concentrations of HDBP were 20 to 30 mg/l and seem to be only slightly increased as compared to the conventional process.

By the presented results we are encouraged to continue our program to design a one cycle Purex process. The new process will be run with high burn-up short cooling time FBR fuel in 1988.

LITERATURE

[1] A.M.Rozen, Atomnaya energiya **7**, 277 (1959)
[2] W.Ochsenfeld, H.Schmieder, S.Theiss, Report KfK 911 (1970)
[3] J.Schön, M. Kluth, H.-J.Bleyl, H.J. Schmidt, this conference
[4] G.Petrich, U.Galla, H.Goldacker, H.Schmieder, Chem.Eng.Sci. **41**, 981 (1986)
[5] S.B.Watson, R.H.Rainey, Report ORNL-TM 5123 (1975)
[6] M.Benedict, T.H.Pigford, Nucl.Chem.Eng., 2nd ed., McGraw-Hill, N.Y. 1981, p. 506
[7] D.E.Benker et al., ENC'86 Transactions, Geneva, June 1-6, 1986, vol.4, p.109
[8] J.Garraway, Extraction'84, ICE Symp.Ser.88, Dounreay 1984
[9] R.Becker, L.Stieglitz, H.Bautz, Report KfK-3836 (1984)

Pyrophoric properties of zircaloy particles from irradiated fuel element claddings

R. Würtz, T. Bereznai, R. Kroebel, H. Andriessen

SUMMARY

In order to assess the fire and explosion hazards of zircaloy fines arising during chopping of fuel elements in a reprocessing plant, artificially produced particles from irradiated and non-irradiated LWR-fuel rod hulls were examined with regard to the pyrophoric properties, using standardized test procedures and devices.

Both, the ignition temperature of a deposited layer of active zircaloy fines (235° C) and of a dust cloud (365° C) were found to be lower than the corresponding values of the non-irradiated material (295° C and 400° C) respectively).

Explosion characteristics indicate that the irradiated fines must be classified in category 2 and the non-irradiated particles in a less dangerous category 1 of a scale of explosion hazards of dusts. Original fines from operational tests with a commercial shear proved to be far less pyrophoric than the artificial material.

Highly active fission products, when mixed with zircaloy fines, merely lowered the ignition temperature of the latter material, if the mixtures were heated up at low rates in a thermoanalytical device (DTA/TG). At higher heating rates (30° C/min), these mixtures produced a vigorous reaction supposedly of the thermite type.

Inertisation with nitrogen proved to be an efficient means for suppressing ignition and explosion of the zircaloy particles.

Dr. R. Würtz, T. Bereznai, Kraftwerk Union AG; Postfach 12 40, D-8757 Karlstein, Federal Republic of Germany
Dr. R. Kroebel, H. Andriessen, Kernforschungszentrum Karlsruhe, Postfach 36 40, D-7500 Karlsruhe, Federal Republic of Germany

1 INTRODUCTION

Fragmentation of light water reactor fuel rods by chopping with a bundle shear in the head-end of a reprocessing plant is always accompanied by the occurance of fine zircaloy particles or even dust, which had given rise to safety considerations because of the known pyrophoric behavior of this alloy.

In order to assess the risks posed by this material to safe operation of the plant, an experimental programme was carried out covering standard-methods for obtaining safety relelvant data for metallic dusts as well as special investigations for determining the oxidation kinetics of zircaloy and the influence of other material, e. e. metallic fission products, on the ignitability /1/.

The program was funded by the Commission of the European Committees under contract No. WAS-246-81-2(D).

Owing to the well-known large differences in the mechanical properties of the as-fabricated zircaloy from the irradiated fuel element cladding material, experiments were conducted with artificially prepared fines of both materials.

The highly active irradiated zircaloy was examined in the hot cells of KWU in accordance with the above mentioned standard methods thus allowing an assessment of the irradiation effect on the pyrophoric properties of zircaloy.

Hazards could arise from zircaloy particles through self ignition of dust layers accumulating in particular areas of the plant or from dust explosions when particles are raised by air flows. The experiments covered both possibilities.

From the operational point of view, three plant areas - shear housing, dissolver and centrifuge or filter - for a potential occurance of zircaloy fines were taken into consideration and will be discussed below.

2 PREPARATION AND CHARACTERIZATION OF THE SAMPLE MATERIAL

For the production of both fines from irradiated zircaloy hulls and non-active cladding tubes, a special device was developed consisting of a rotating disk file with a pneumatic feed for the tube material. To prevent the filings from spontaneous ignition, occasionally occuring when preparing sample material in air, the apparatus was inerted with argon.

Due to the increased brittleness of the irradiated zircaloy hulls, the filings of this material differed substantially in shape and size from the inactive fines. A visual impression on the differences is given in fig. 1, presenting scanning electron microscope images of both materials. Consequently, the active material had a larger part of fine particles as shown in fig. 2 for a sieve fraction < 100 µm.

Fig. 1: SEM examinations of irradiated (top) and non-irradiated (bottom) zircaloy fines (< 100 µm)

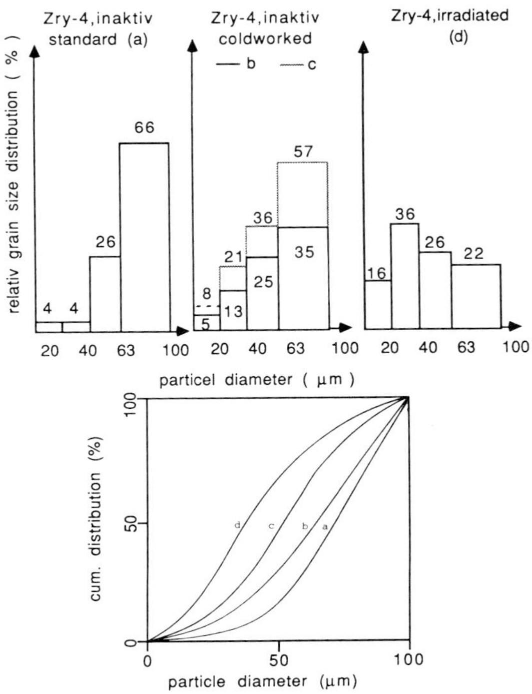

Fig. 2: Grain size distribution of different zircaloy fines (fraction < 100 µm)

In order to produce non-irradiated fines with a similar grain size distribution as the active fines, a special type of cold-worked tubes was used as sample material.

As shown in fig. 2, the grainsize distribution of this material was between the irradiated and the standard non-irradiated zircaloy, proving the influence of brittleness on the grain size.

3 IGNITABILITY OF ZIRCALOY DUST LAYERS

Following the recommendation of the "Bundesanstalt für Materialforschung und -prüfung (BAM)", qualified for the assessment of fire hazards, two standard test methods for the characterization of the pyrophoric properties of zircaloy dust layers were applied: ignition on a heated plate and self ignition in an oven (Gliwitzky oven) /2/, /3/.

The results of the ignition temperature determinations on the heated plate given in table 1 show a marked decrease of the ignition temperature with decreasing grain size. Corresponding measurements with the same inactive sample material, carried out by BAM, yielded very similar results (table 1) proving the reproducibility of the test procedure.

material	fraction (µm)	ignition temperature $T_{ign.}$ (°C) KWU	BAM	$T_{non\ ign.}$* (°C)
unirradiated	<100	320	330	316
	<200	335	---	330
	100-500	380	---	376
	200-500	395	---	390
	400-500	400	415	395
	>500	415		410
irradiated	1) <100	236	---	231
irradiated	2) <100	265	---	260
irradiated	3) <100	256	---	250

*) maximum temperature of three succeeding experiments without ignition

1) material from Zry-4-hulls,with fuel dissolved with HNO_3

2) material from Zry-4-hulls,with fuel mechanically removed

3) material from Zry-2-hulls,with fuel mechanically removed

Table 1: Ignition temperatures of zircaloy dust layers

The experiments with active material were confined to a sieve fraction less than 100 µm, which has proved to cover the worst case in respect to ignition.

The ignition temperature of the active zircaloy fines - approximately 250° C - is significantly lower than that of the unirradiated material. This is probably due to the higher proportion of very fine particles in the active filings and not to primary irradiation effects. This is also supported by examinations of the oxidation kinetics of both materials (to be published later) showing a very similar reaction rate for irradiated and unirradiated zircaloy. No major differences of ignition temperatures of zircaloy2 and zircaloy4 have been found nor has there been detected an influence or preconditioning of the zircaloy hulls before filling: part of the irradiated fuel rod sections were boiled in nitric acid in order to dissolve the UO_2-fuel whereas the fuel was removed by pushing it out of the cladding mechanically in other samples.

A similar relation between irradiated and unirradiated fines has been found for the self ignition temperature in the Gliwitzky oven. The results of these examinations are given in figure 3 which shows the dependence of self ignition temperature, T_S, on the volume to surface ratio, v/s, of the fines, heated up in cylindrical wire baskets.

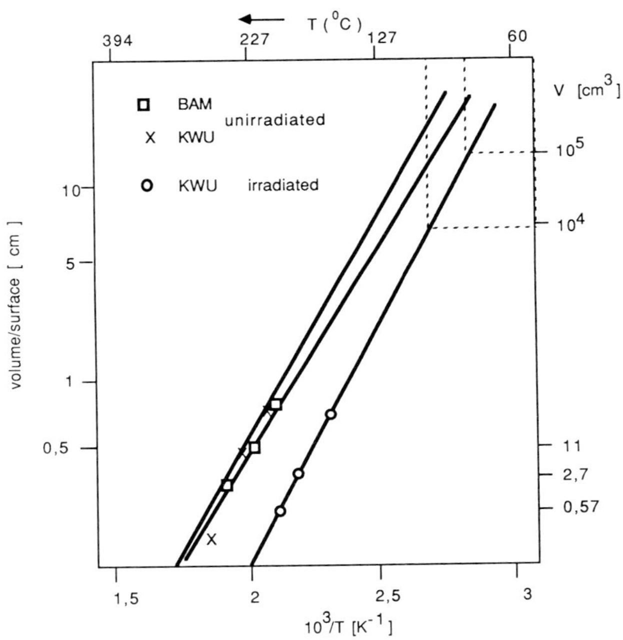

Fig. 3: Dependence of self-ignition temperature of zircaloy fines (< 100 µm) on the volume/surface ratio

According to a semi-empirical approach proposed by BAM /2/, there is generally a linear relation between $1/T_S$ and v/s which was also found in this work (fig. 3). Therefore, by extrapolating the straight line in fig. 3, one can estimate that, for instance, ignition below 100° C should not be expected for samples as large as 10 litres. As regards the inactive fines, the sample must be even larger by a factor of about 5. Again, we assume that this difference is mainly due to shape and size of the particles and not to irradiation defects.

As the occurence of zircaloy fines at reprocessing may accasionally be accompanied with insoluble residues of fission products (IFP) which are highly radioactive and therefore heat producing, mixtures of zircaloy fines and genuine fission product residues from the Karlsruhe Reprocessing Plant (WAK) have been investigated by thermal analyses.

These mixtures exhibited different behaviour when heated up at different rates in the thermoanalytical device in which it was possible to measure the weight gain (TG) as well as the temperature rise in the sample relating to an inert sample (DTA).

At low heating rates (5° C/min) the fission products only acted as an inert diluting material for the zircaloy fines, approaching the reaction rates of the pure zircaloy with decreasing content (fig.4). The weight gain observed at the end of the experiments corresponds in all cases to the total oxidation of the zircaloy in the samples.

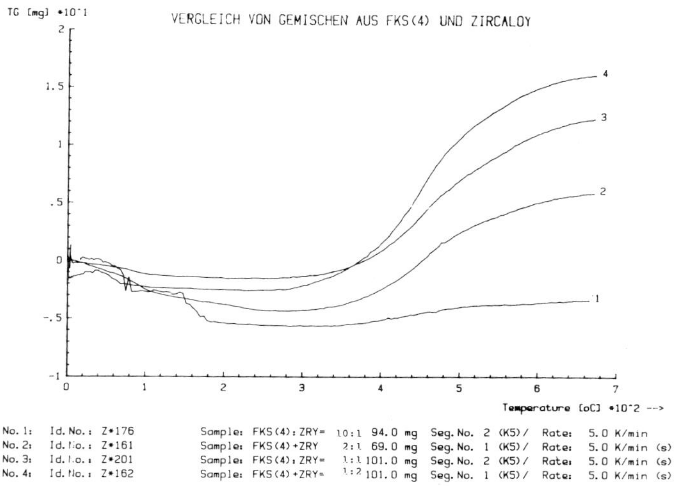

Fig. 4: Oxidation of zircaloy/fission product residue mixtures at low heating rates (5° C)

At high heating rates (30° C/min) a spontaneous heat releasing reaction occurs (fig.5) and generally part of the material is thrown out of the furnace, producing a sharp decrease in the sample weight.

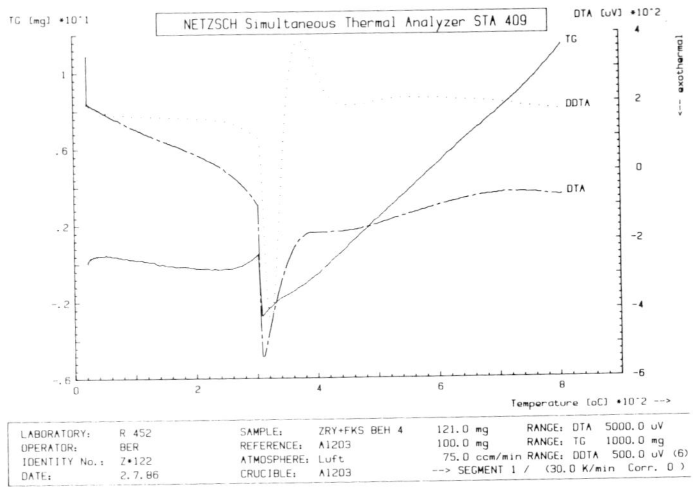

Fig. 5: Oxidation of zircaloy/fission product residue mixtures at high heating rates (30° C)

This suggests a thermite type reaction of zircaloy fines with fission product oxides like MoO_3.

4 EXAMINATION OF IGNITION AND EXPLOSION OF DUST CLOUDS

Although dust clouds can ignite as a result, for instance, of electric sparks or mechanical impact, assessment of the ignitability is performed using a standardized method, determining the ignition temperature in the Godbert-Greenwald Apparatus /3/ shown in fig. 6.

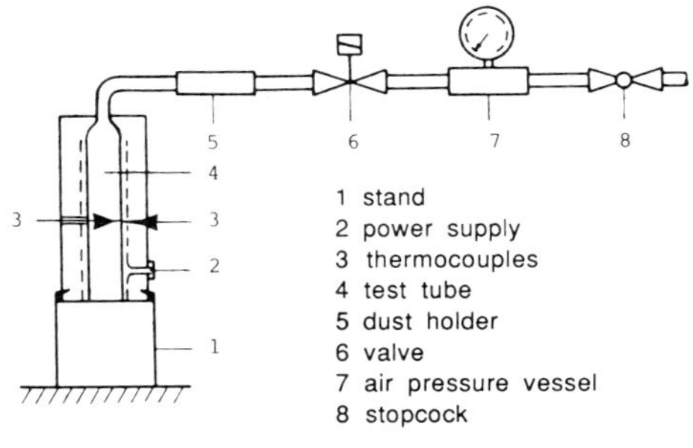

Fig. 6: Godbert-Greenwald furnace for the determination of the ignition temperature

A sample of zircaloy fines is blown into a horizontal furnace, preheated to test temperature. The lowest temperature at which an ignition occurs is defined as ignition temperature. At the usual overpressure of 0.3 bar for transferring the sample into the furnace, zircaloy fines generally ignited even at room temperature due to the impact of particles on the wall, showing their extreme sensitivity to mechanical impact. The pressure for blowing in the sample therefore had to be lowered to 0.2 bar to ensure suitable performance of the measurement. The furnace temperature had been measured in steps of 5° C until ignition occurred.

The results given in table 2 show good agreement between KWU and BAM for the unirradiated material and again a somewhat lower ignition temperature of the irradiated zircaloy fines. The decreasing ignition temperature at smaller sieve fractions of fines like the former results suggests that the grain size distribution is the major property influencing pyrophoricity.

material	fraction size	KWU				BAM	
		T ign.	sample weight	T non ign.	sample weight	T ign.	sample weight
	[µm]	[°C]	[g]	[°C]	[g]	[°C]	[g]
unirradiated	<100	~400*	1,0		0,8	390	0,8
	100-500	451	0,5	446	0,8	-	
	200-500	450	0,5	445	0,8	450	0,8
	400-500	470	0,5	465	0,8	460	0,8
	>500	465	0,5	460	0,5	-	
	100-400	-		-		440	0,8
irradiated	<100	365	0,7	360	0,5-1,0		

Table 2: Ignition temperature of irradiated and non-irradiated zircaloy fines dispersed in air

The protective measures to be taken if there exists a risk for a dust explosion are generally determined by the following characteristics: maximum explosion pressure, maximum rate of pressure rise and limiting oxygen for explosion.

Maximum explosion pressure and pressure rate are determined by means of a special device developed for conventional safety techniques (fig. 7). It consists principally of a stainless steel sphere of 20 l volume in which the particles are homogeneously distributed by being blown in through a specially shaped valve.

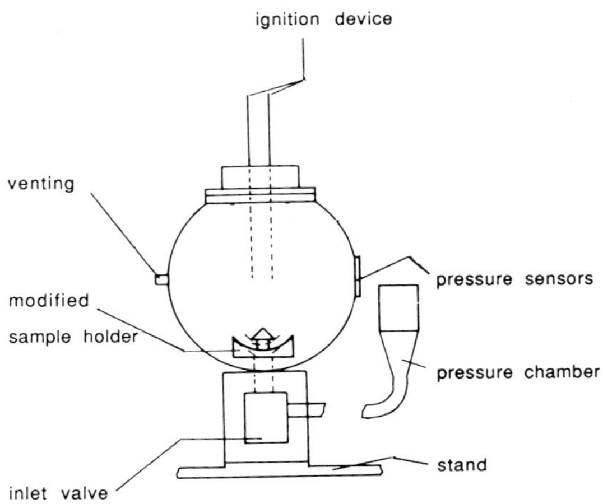

Fig. 7: Explosion chamber (20 l) for measurement of explosion characteristics

Usually, the cloud is then ignited by a pyrotechnical primer and explosion pressure and pressure rate are measured using piezo-electric transducers and computer-controlled high-speed electronic recording equipment /5/, /6/.

In the case of zircaloy fines premature ignition due to mechanical impact of the particles had always been observed so that the pyrotechnical ignitor was not required and could be omitted.

Experience /4/ shows that both explosion pressure and rate of pressure rise approach boundary values when the dust concentration is continuously increased. These maximum values are used as characterizing data for the material. Fig. 8 shows the results for unirradiated and irradiated zircaloy fines (less than 100 µm) indicating that the maximum values are achieved with dust concentrations above 4 kg/m^3. Maximum explosion pressure of both materials differ only slightly (7.6 bar for irradiated and 6.2 bar for unirradiated zircaloy fines less than 100 µm).

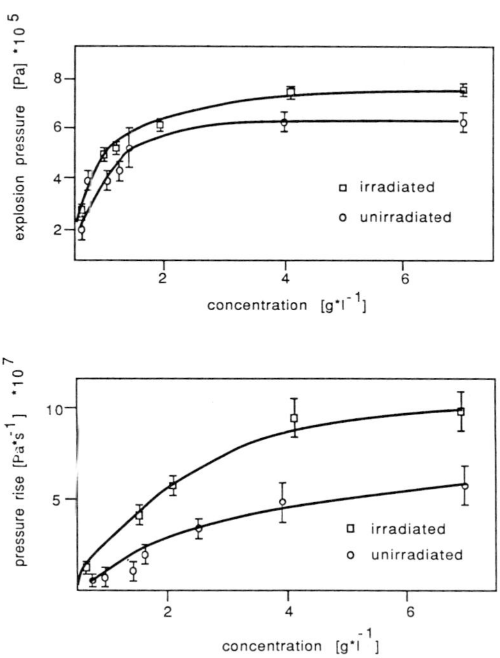

Fig. 8: Dependence of explosion pressure and pressure rise on zircaloy dust concentration

The rate of pressure rise of the irradiated material, however, exceeds that of the unirradiated zircaloy fines by a factor of approximately two, classifying the irradiated material in category 2 and the unirradiated material in a less dangerous category 1 on a scale of explosion hazards of dusts.

The reason for the higher rate of pressure rise of the irradiated zircaloy is supposed to be related to the large quantity of fine particles in the fraction used, which was only limited by the upper boundary of 100 um. Smaller particles apparently react much faster than larger ones, some of which were left unreacted after the test.

Inertisation was experimentally examined in the 20 l-sphere by BAM using N_2 and Ar, with the results given in table 3. Surprisingly, N_2 proved to be the more efficient inerting media since a chemical reaction between N_2 and Zr generally occurs at elevated temperatures. As can be seen from table 3, dust explosions could be observed even with the lowest O_2 contents but the maximum pressure and rate of pressure rise are decreasing to insignificant values with increasing N_2 content.

volume fraction O_2 (%)	volume fraction and type of inerting gase (%)		maximum pressure rise (bar s^{-1})	explosion pressure (bar)
21	79	N_2	69	3,9
21	79	Ar	175	4,5
15	85	N_2	15	3,4
15	85	Ar	40	3,6
10	90	N_2	10	2,9
10	90	Ar	30	3,4
5	95	N_2	5	1,6
5	95	Ar	20	3,0
1	99	N_2	-	1,0
1	99	Ar	5	1,1

Table 3: Results of inertisation tests

5 IMPLICATIONS ON PLANT DESIGN AND OPERATION

Owing to the requirements by BAM to use the fines fraction of < 100 µm for most of the characterisation tests, the results described above are certainly very conservative or even pessimistic concerning the reactions with air or oxygen. Shearing tests /7/ revealed a particle size distribution with far fewer fines below 100 um and mainly particles in the mm size range. Our own preliminary tests on inactive shear fines with the TG method indicate a temperature range of 600 °C for ignition, which is about 200 °C higher than for our standard fraction < 100 µm.

Furthermore, the British tests /7/ revealed another positive effect within the shear which suppresses dust cloud ignition, namely the high amount of inerting UO_2-dust present diluting the zircaloy particles by a factor of 10 to 100. These two realistic circumstances are apparently responsible for the fact that zircaloy fines explosions within the shear have - up to now - not been reported and are rather unlikely to happen.

Nevertheless, the shear design should not allow accumulation of zircaloy fines in the range above several hundred grams, or if this cannot be avoided, nitrogen inertisation should be provided, which according to the afore-mentioned results practically prevents hazardous conditions in the shear.

The second area of possible zircaloy fines accumulation is the dissolver itself or any place to which fines can be transported in slurry. At these places, dust clouds are unlikely, but ignition cannot be excluded. These ignitions are harmless as long as the amount of fine material is restricted. Quantitative assumptions have to be made for the safety case and care has to be taken that inspection and removal of particle accumulations can be carried out safely.

The third place of possible accumulation of fines is the feed clarification stage by centrifuge or filter. Here our results from thermogravimetric measurements show that self-heating or heat input from external sources should be slow, as heating rates of 5 °C/min merely resulted in a weight increase without violent reaction, whereas heating rates of 30° C/min produced an intensive reaction, probably of the thermite type, between zircaloy and fission product oxides, e. g. MoO_3.

Generally, a low ratio of zircaloy to the bulk of undissolved residues is favourable with regard to safety considerations. Final conclusions in each specific safety case can only be drawn after careful evaluation of the design information and therefore may differ from plant to plant. But in spite of the facts reported on pure zircaloy fines of low particle size, incidents can be avoided by proper design combined with operating precautions.

Further investigations will be undertaken at TH Darmstadt (Prof. Münzel) to study the behaviour of zircaloy fines with liquid and solid admixtures which could be relevant when dealing with the conditioning of fines for disposal.

LITERATURE

/1/ H. Andriessen, R. Kroebel, T. Bereznai, R. Würtz, H. Hattwig, W. Hensel, R. Oßwald
Untersuchungen zum Brand- und Explosionsverhalten von Zircaloy-Feinteilen
EUR-11120, 1987

/2/ G. Leuschke, R. Oßwald
"Bedeutung und Ermittlung von sicherheitstechnischen Größen brennbarer Stäube"
VDI-Berichte Nr. 304, 1978

/3/ G. Leuschke
Über die Untersuchung brennbare Stäube auf Brand- und Explosions gefahren
Staub 26 (1966) 2, Seite 49/57

/4/ H. G. Dorsett, M. Jacobson, J. Nagy and R. P. Williams
Laboratory Equipment and Test Procedures for Evaluating Explosibility of Dusts
US Bureau of Mines, Report of Investigation 5624 (1960)

/5/ W. Bartknecht
Explosionen - Ablauf und Schutzmaßnahmen Springer-Verlag Berlin 1978

/6/ P. Field
Dust Explosions
Elsevier Scientific Publishing Company
Amsterdam - Oxford - New York 1982

/7/ N. J. James et al.
Zircaloy Hazards in Nuclear Fuel Reprocessing
I. Chem. E. Symposium Series No. 97, 1985

Is solvent radiolysis a safety problem for PUREX plant performance?

A.H. Stollenwerk, M. Weishaupt, L. Stieglitz, R. Becker

SUMMARY

During two campaigns with shortly cooled low burn-up and with high burn-up fuel of 2 to 4 years cooling time, respectively, the solvent quality of the first solvent cycle of the WAK plant has been analyzed. Quick analytical methods were used to obtain information on the concentration of radiolysis products.

Applying formation rates from a solvent radiolysis model, the total radiation absorbed dose of the solvent was estimated and could be broken down in to the α- and β/γ -contributions. These figures compare well with those obtained if one considers the irradiation history of WAK plant solvent.

Thus, β/γ -irradiation in the first extraction bank accounts for the observed β/γ -absorbed dose. The α-irradiation of extracted plutonium contributes the major part of radiation damage to the solvent. However, only one half of the observed α-dose is due to the stay of plutonium loaded solvent in the extraction banks. The other half is caused by α-irradiation of organic slip phase, which is collected in aqueous plutonium product catch tanks and is discontinuously fed back to the solvent cycle. Diluent washing will be provided for aqueous plutonium product streams in a future commercial reprocessing plant. Hence, its solvent will accumulate α-radiation damage to a lesser extent than the solvent of the WAK plant.

Dr. A. H. Stollenwerk, Dr. M. Weishaupt, Wiederaufarbeitungsanlage Karlsruhe Betriebsgesellschaft mbH, Postfach 2 20, D-7514 Eggenstein-Leopoldshafen 2, Federal Republic of Germany
Dr. L. Stieglitz, Dr. R. Becker, Institut für Heisse Chemie, Kernforschungszentrum Karlsruhe, Postfach 36 40, D-7500 Karlsruhe, Federal Republic of Germany

1 INTRODUCTION

The selective and moderate affinity of tributyl-phosphate (TBP) to complexing uranium (IV, VI) and plutonium (IV) allows the separation of these species from fission products with high yield and extreme product purity. It is a multi-stage, counter-current extraction process in which TBP, diluted in a nonpolar hydrocarbon, is vigorously mixed with an aqueous nitric acid solution of dissolved fuel. The process places a high radiolytic load on the solvent from β/γ - radiation of the fission products. In addition, the total α-decay energy of extracted plutonium will be absorbed.

The TBP/dodecane solvent is known to be extremely radiation resistant - in fact, besides its complexing qualities, this has been the most important criterion in choosing the PUREX process solvent. However, there is radiolytic production of a variety of derivatives of TBP and dodecane in small yield. These may be characterized according to their effect on the process performance, as follows:

- formation of hydrogen and volatile hydrocarbons. There are no safety problems associated with these products (1).

- The formation of dibutyl and monobutyl phosphate by fission of butyl groups from TBP. These radicals are also formed by hydrolysis. Formation rates are well known from laboratory experiments (2) and in accord with those found in the technical process (3). Acidic products are removed from the solvent during alkaline wash treatment.

- Formation of products with higher molecular weight than those of TBP or dodecane by reactions of radiochemically activated TBP or dodecane, or parts thereof.

The latter products show several troublesome properties: some of them are either not or rather incompletely removed from the solvent during normal alkaline treatment, hence they accumulate on solvent recycling. Some of them show a complexing affinity towards fission products or plutonium/uranium, and hence deteriorate product purity and enhance product losses. Some of them are surfactants which may significantly extend the phase separation time. Above all, the phase separation of emulsions formed on alkaline treatment of solvent will be affected.

Our paper will review the radiochemical model of formation and analytical methods to determine the concentration of these so-called "do-bads". Rates of their formation will be given, both for α- and β/γ-irradiation of the solvent. Based on the analytical determination of contaminants of irradiated solvent samples, a method is given to estimate the total absorbed α- and β/γ-radiation dose.

The method has been applied to the solvent of the Karlsruhe reprocessing pilot plant (WAK) during two campaigns with rather different kinds of spent fuel: in one campaign shortly cooled low-burn-up fuel, in the other high-burn-up fuel being cooled for 2 to 4 years was processed.

Total absorbed dose estimates have been verified by calculating the irradiation history of the solvent during these campaigns. Using readily accessible data on spent fuel, plant chemical flow sheet and lay-out of extraction equipment, the total absorbed dose of solvent per pass through the first extraction cycle has been calculated. The total number of passes through the cycle has been derived from a balance of daily solvent losses.

Solvent is continuously lost from the cycle by entrainment in aqueous raffinates or radiolytical or chemical degradation and is discontinuously replaced by fresh solvent.

Finally, the maximum absorbed dose will be estimated for the solvent of a future commercial spent fuel reprocessing plant.

2 CHEMICAL MODEL OF SOLVENT RADIOLYSIS, ANALYTICAL METHODS

The quality of the PUREX process solvent may be established by testing its performance in the reversible extraction of plutonium and the extraction of zirconium. These solvent properties are closely related to its absorbed dose. The chemical nature of the species responsible for them has been asserted and, as a direct analysis of one single product is not sufficient, methods have been developed for the determination of different classes of compounds. These methods are sensitive, easy to handle, and quick (4).

Four classes of compounds were found to characterize the solvent quality sufficiently:

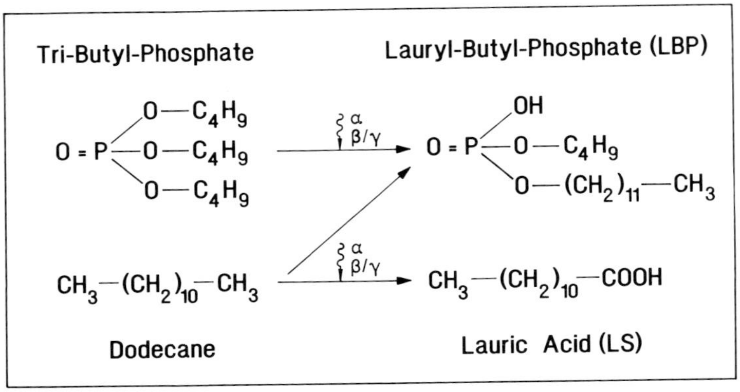

Fig. 1: Radiochemical build-up of surfactants

- Acidic esters of phosphoric acid, which may be produced by scission of one butyl group of TBP and replacement of another by a long chain alkyl group. The model compound is lauryl-butyl-phosphate (LBP) (fig. 1).

- Alkyl carbonic acids, which are produced on oxidation of diluent constituents. The model compound is lauric-acid (LS) (fig. 1).

- Nitro alkanes, which may be produced on nitration of diluent constituents with nitrous oxide. The model compound is nitrododecane/nitroalkane (NA).

- Alkyl nitrates, which may be produced on reaction of radiolytically formed butene with nitric acid. The model compound is butylnitrate/alkylnitrate (AN).

Nitroalkanes and alkylnitrates are not active surfactants. However, they may serve as dose indicator in irradiated solvent. Since they may also be formed thermochemically, they are semiquantitive dose indicators only.

A solvent sample is first washed with sodium carbonate to remove nitric acid and traces of heavy metals. The quality test then comprises several steps (cf. (4) and fig. 2):

LBP	Conductivity / $NaHCO_3$
LS+LBP	Conductivity / Na_2CO_3 Surface tension / Na_2CO_3
$R-NO_2$ (NA)	Polarography
$R-ONO_2$ (AN)	Polarography

Fig. 2: Analytical determination of radiolysis product concentration

- the conductivity measurement of a sample equilibrated with sodium bicarbonate determines only the content of the lauryl-butyl-phosphate-type (LBP-type) compounds;

- the conductivity measurement of a sample equilibrated with sodium carbonate, determines the content of LBP-type and lauric acid-type (LS-type) compounds because the salts of both types are formed;

- measurement of the interfacial tension of solvent and sodium carbonate solutions reflects the combined surface activity of LBP- and LS-type compounds;

- polarographic determination of nitroalkane- and alkylnitrate-type (NA- and AN-type) compounds, respectively.

In doing so, information is obtained on the concentrations of the main contaminants impairing solvent performance as well as on the total absorbed α- and β/γ -radiation dose.

3 EVALUATION OF THE α- AND β/γ -RADIATION ABSORBED DOSE

Since the formation of solvent radiolysis products has been shown to be linearly dependent on the radiation absorbed dose with different rates for α- and β/γ -radiation, respectivly (cf. (4) and fig. 3), the evaluation of α- and β/γ -radiation absorbed dose from measured radiolysis product concentrations seems to be straightforward. Because of experimental uncertainties, however, we prefer an optimization proedure which also takes into account the measured surface tension.

	α	β/γ
LBP	$8 \cdot 10^{-6}$	$3.7 \cdot 10^{-6}$
LS	$1.6 \cdot 10^{-5}$	$3.4 \cdot 10^{-5}$
$R\text{-}NO_2$ (NA)	$9.7 \cdot 10^{-5}$	$6.3 \cdot 10^{-5}$
$R\text{-}ONO_2$ (AN)	$3.9 \cdot 10^{-4}$	$2.7 \cdot 10^{-4}$

Fig. 3: Formation rates

On increasing the concentration of either the LBP or LS model compound in TBP/dodecane, an exponential decrease of the surface tension is expected. Beyond a critical concentration, the solvent/aqueous surface will be loaded with surfactant and the surface tension will not drop below a residual value (fig. 4).

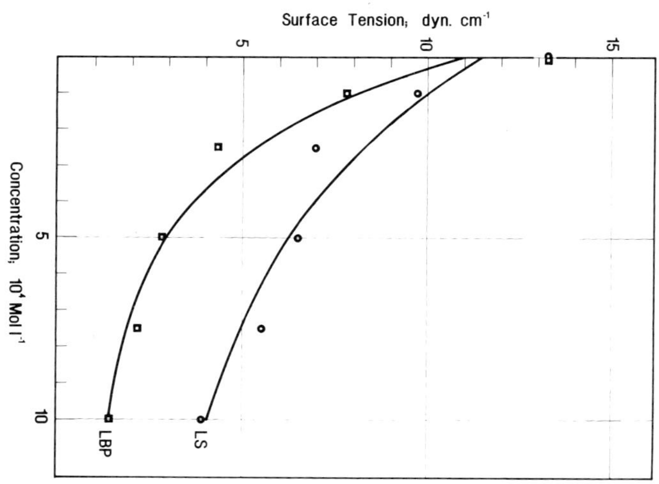

Fig. 4: Surface tension of TBP/Na_2CO_3 vs. concentration of lauryl-butyl-phosphate and lauric acid

Parameters to fit the exponential dependence of surface tension (s) on concentration (c, cf.fig. 5) are listed in table 1.

$$S = S_k + S_0 \exp(kc)$$

$$S = S_{1k}c'_1 + S_{2k}c'_2 + $$
$$+ (S_{10}c'_1 + S_{20}c'_2)\exp(K_1c_1c'_1 + K_2c_2c'_2)$$

$$c'_1 = \frac{c_1}{c_1 + c_2} \qquad c'_2 = \frac{c_2}{c_1 + c_2}$$

Fig. 5: Radiolysis product concentration and surface tension

As can be seen from figure 4, the measured tension of the original solvent is somewhat underestimated by our exponential formula.

Compound	S_K(dyn/cm)	S_O(dyn/cm)	k(l/mol)
LBP	1	10	- 3300
LS	2,5	9	- 1750

Table 1: Parameters for calculation of surface tension of TBP/Na_2CO_3 contaminated with LBP or LS

The combined influence of both model compounds, LBP and LS, on the solvent/Na_2CO_3 surface tension may be calculated by superimposing the exponential contributions of both compounds (cf. fig. 5) using parameters of table 1.

This assmuption will now be checked:

In two experimental series, TBP/dodecane was irradiated using an external Co-source and plutonium with high Pu-238 content, respectively. Using the formation rates for radiolysis products of figure 3, the concentration dependent surface tension of solvent/Na_2CO_3 may be calculated and compared to experimental values of either series. Figure 6 shows the results obtained with the α-irradiated solvent; except for the underestimation of the initial solvent's surface tension the fit of the experimental data seems satisfactory.

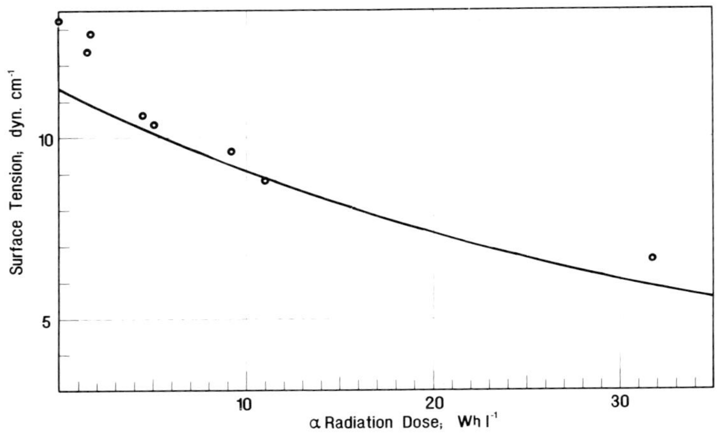

Fig. 6: Surface tension of Solvent/Na_2CO_3 vs. absorbed α-radiation dose

Figure 7 shows the results obtained with the β/γ-irradiated solvent. The calculated surface tension falls short of the experimental data by 2 dyn/cm. The measurement will be repeated up to a higher absorbed dose range.

If the concentrations of LBP, LS, NA, AN, and the surface tension of an irradiated solvent sample are given, it is possible to estimate the α- and β/γ-radiation absorbed dose.

Using the formation rates of figure 3 and the surface tension formula of figure 5, one can calculate the concentratons of the radiolysis products and the surface tension and can compare these data with the experimental findings. The α- and β/γ-radiation absorbed dose figures are then varied until convergence is found between calculated and experimental data.

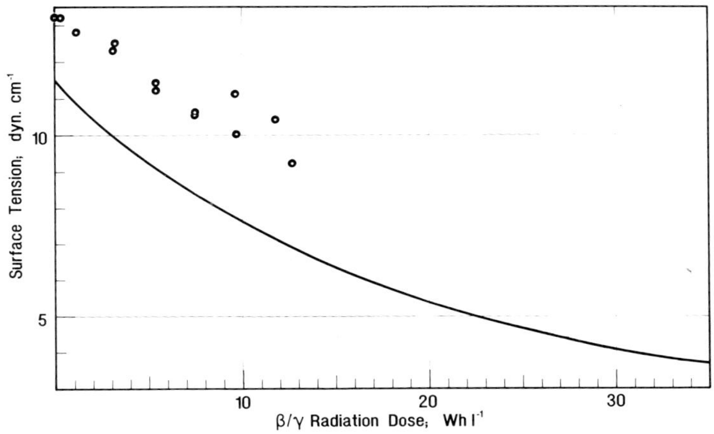

Fig. 7: Surface tension of solvent/Na_2CO_3 vs. absorbed ß/γ -radiation dose

4 QUALITY TESTS OF WAK PLANT SOLVENT

The quality of the solvent from the first extraction cycle of the Karlsruhe reprocessing pilot plant (WAK) is tested on a routine basis. The results of test series during two campaigns with rather different kinds of spent fuel will be discussed.

In the first campaign (6. MZFR) fuel with a burn-up of 13 GWd/t and cooled for 1 to 4 years was processed, and in the second campaign (5. KWO) average fuel burn-up rated 30 GWd/t with a cooling time of 2 to 4 years.

Obviously, the concentrations of all solvent radiolysis products increased during the second campaign, while the surface tension of solvent to sodium carbonate solution decreased.

The α- and ß/γ-radiation absorbed dose was estimated from these data. Several pairs of α- and ß/γ-radiation absorbed doses are given in table 3 along with the quality indicators to be associated with each pair.

Table 2 lists average indicator concentrations of 1st cycle solvent during these campaigns:

Campaign	LBP	LS	NA	AN	S
	<----	(mMol/l)		---->	(dyn/cm)
6. MZFR	0.162	0.413	3.9	14	6.67
5. KWO	0.226	0.476	4.0	16	5.79

Table 2: Quality indicators for first cycle solvent of WAK during two reprocessing campaigns

From inspection of table 3, we conclude that 20 Wh/l α- and 3 Wh/l ß/γ-radiation absorbed dose will best account for the observed quality indicators of the first cycle solvent of WAK. Hence the α-radiation of extracted plutonium causes more than 75 % of the radiation damage to the solvent. Contributing only less to the radiation damage of the

D	$D_{ß/γ}$	LBP	LS	NA	AN	S
<--(WH/l)-->		<------	(mMol/l)	------>		(dyn/cm)
20	3.0	0.171	0.422	2.1	8.6	6.73
17	6.0	0.158	0.476	2.0	8.3	6.47
23	0	0.184	0.368	2.2	9.0	6.96
0	23	0.085	0.782	1.5	6.2	4.93
Observed		0.162	0.413	3.9	14	6.67

Table 3: Quality indicators of first cycle solvent of WAK calculated from α- and ß/γ-radiation absorbed dose and compared to experimental figures

solvent, the β/γ-radiation absorbed dose is poorly estimated from solvent quality indicators. The irradiation accounts for two thirds of nitroalcane and alcylnitrate concentration in the plant solvent, one third is presumably produced by thermochemical attack of nitrous oxide/nitric acid to the solvent.

The best fit radiation absorbed dose pairs for WAK first cycle solvent of two different campaigns are given in table 4.

Campaign	D_α	$D_{\beta/\gamma}$
	<--- (Wh/l)	--->
6. MZFR	20	3.0
5. KWC	28	1.2

Table 4: Radiation absorbed dose of first cycle solvent of WAK during two reprocessing campaigns

Obviously, a higher plutonium throughput with considerably higher Pu-238 content, to be associated with processing of high burn-up fuel, caused a significant increase of α-irradiation of the plant solvent. The apparent decrease of β/γ-radiation absorbed dose must be considered meaningless because it is within the uncertainty of this estimation.

Solvent quality indicators thus give a plausible estimate of the radiation absorbed dose. Quantitative verification will be obtained from a detailed consideration of the solvent's irradiation history.

5 IRRADIATION HISTORY OF WAK PLANT SOLVENT

The calculation of a solvent's irradiation history is divided into several sub-tasks:

- from campaign specific fuel data (burn-up, cooling time, etc.) ß/γ-dose rates of highly active plant feed solutions are evaluated as well as specific α- and ß-dose rates of fuel plutonium.

- from readily accessible plant data (chemical flow sheet, campaign specific throughput, layout of extraction equipment, etc.) total radiation absorbed dose of the solvent per pass through the first extraction cycle is calculated.

- from a balance of daily solvent losses, the total number of solvent passes through the first extraction cycle is derived. This relation of solvent flow to continuous solvent loss is called recycling factor.

We shall conclude with an estimation of the total absorbed dose of solvent from irradiation history, i. e., absorbed dose per pass times recycling factor.

5.1 CAMPAIGN SPECIFIC DOSE RATES

The concentrations of most of the radioactive fission products in highly active feed solution of the WAK plant have been obtained from a routine analysis of the plant feed. Some nuclides (e. g. Sr/Y-90) had to be supplemented from burn-up calculations using the KORIGEN code (5). The nuclide decay heat has been split into a ß- and a γ-part, respectively, using a table of decay schemes (6) and a catalogue of photon spectra (7). The resulting dose rates are listed in table 5.

Fuel Solution	DR_{β} <--- (W/l)	DR_{γ} --->
WAK 6. MZFR	0,16	0,054
WAK 5. KWO	0,28	0,18
FCP Uranium FE	0,20	0,19

Table 5: ß- and γ-dose rates of representative fuel solutions of the WAK plant and of a future commercial plant (FCP)

For sake of comparison, ß- and γ-dose rates of highly active feed solution expected for a future commercial plant have been added to the table. These were obtained using a burn-up calculation for uranium fuel with a burn-up of 40 GWd/t and 7 years cooling time, data which are thought to be representative for the fuel of a future commercial plant.

The plutonium element concentration of fuel solutions and its plutonium isotopic composition are routinely analysed at WAK. Since WAK has processed fuel with a burn-up of 40 GWd/t, the plutonium content and the plutonium isotopic composition of fuel which will be processed at the Wackersdorf site are well known. Considering the decay schemes of plutonium isotopes (6, 7), the decay heat of plutonium from different fuel elements is broken down to radiation specific dose rates (table 6).

Fuel Plutonium	DR_{α} <-----	DR_{β} (mW/g)	DR_{γ} ----->
WAK 6. MZFR	5,53	0,4	$1{,}44 \times 10^{-3}$
WAK 5. KWO	10,8	0,695	$4{,}81 \times 10^{-3}$
FCP Uranium FE	13,9	0,621	$6{,}38 \times 10^{-3}$

Table 6: Specific dose rates of plutonium from representative fuel of the WAK plant and of a future commercial plant (FCP)

Particularly for high burn-up fuel, the plutonium-238 contribution will dominate the α-dose rate. Plutonium-241 always contributes the major part to the ß-dose rate. The plutonium γ-dose rates are low enough to be omitted from considering the radiation damage to the solvent.

5.2 ABSORBED DOSE PER PASS THROUGH THE FIRST EXTRACTION CYCLE

The ß/γ-radiation of fission products will be absorbed by the solvent in the dispersed phase of the first and highly active extraction bank only. Figure 8 shows the appropriate expression (8) to calculate the ß/γ-radiation absorbed dose, using the average time of the solvent's stay in the first extraction bank (t), the dose rates from table 5, and the average hold-up of the dispersed phase (ε). 50 % of - γ- radiation are thought to be taken up by the solvent. The solvent flow through the first extraction bank (F_{HAX}) is related to the total solvent flow through the first extraction cycle (F_{1CW}). Fission products are diluted, hence the dose rates are reduced by the scrub solution from up-stream fission product scrubbers which is fed into the first extraction bank together with the fuel solution, the appropriate factor being 3/5.

$$D_{\beta/\gamma} = t \cdot (DR_{\beta} + 0.5\, DR_{\gamma}) \cdot \varepsilon \cdot F_{HAX} / F_{1CW}$$

$$D_{\alpha/\beta} = DR_{Pu} \cdot \sum_{i} \cdot t_i \cdot C_{i\,Pu} \cdot F_i / F_{1CW}$$

Fig. 8: Calculation of absorbed dose per pass through the first extraction cycle

Depending on the processed fuel, β/γ-radiation absorbed doses per pass of solvent through the first WAK extraction cycle are obtained:

6. MZFR 0,010 Wh/l
5. KWO 0,020 Wh/l

To calculate the α- (and β-) dose uptake of the solvent, caused by the radiation of extracted plutonium, each extraction bank of the first solvent cycle has to be considered separately, the appropriate expression being shown in figure 8.

The α- and β-dose rates are taken from table 6. Extractor specific data to be used are the average time of the solvent stay (t), the average plutonium concentration in the solvent (C_{Pu}), and the solvent flow (F), respectively. The absorbed dose is related to the total solvent flow through the first WAK solvent cycle (F1CW).

The first solvent cycle of WAK comprises two extraction cycles as shown in figure 9: the first extraction cycle providing for uranium/plutonium coextraction and fission product scrubbing (HA/HS), plutonium stripping and washing (1BX/1BS), uranium stripping (1C), and solvent scrubbing (SS); the second plutonium cycle providing for plutonium extraction and scrubbing (2A/2AS), plutonium stripping (2B), and return of used solvent to the first cycle via buffer and 1BS-bank.

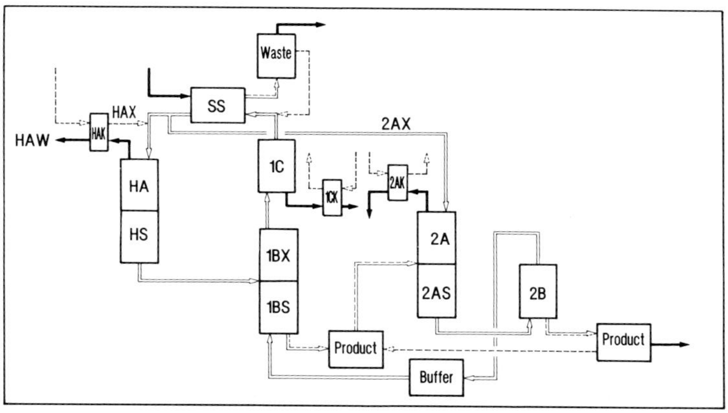

Fig. 9: The first solvent cycle of the WAK

Traces of organic phase will usually be entrained or dissolved in aqueous effluents of extraction banks. Separate diluent washing stages are therefore provided for highly-active aqueous raffinate (HAK), aqueous raffinate of the 2nd plutonium cycle (2AK), and uranium product (1CK)(cf. fig. 9), however, neither the plutonium product of the first nor of the second cycle are washed. Organic phase will therefore accumulate in product catch tanks of the first and second cycles, respectively. Organic hold-up of the second cycle product is discontinuously recycled to the first cycle product to be fed discontinuously to the plutonium extraction bank.

The solvent radiolysis in the organic slip phase has to be taken into account, using reasonable estimates of the entrained amount, of average residence time in the product catch tanks, and of the plutonium concentration, respectively.

In fact, about 50 % of α-radiation absorbed dose per pass of solvent flow through the first solvent cycle has to be attributed to solvent radiolysis "on entrainment path" (see table 7).

Campaign	D_α <--- (Wh/l)	$D_{\beta/\gamma}$ --->
6. MZFR	0,024	0,012
5. KWO	0,067	0,024

Table 7: Radiation absorbed dose per pass of solvent through the first WAK solvent cycle during two reprocessing campaingns

5.3 SOLVENT LOSSES, RECYCLING FACTOR

Being entrained or dissolved in aqueous raffinates or products and being radiolytically or chemically degraded, solvent is continuously lost from the solvent cycle and is discontinuously replenished with fresh solvent (cf. fig. 10). Thus the solvent residence time in a cycle is limited which may be expressed as total number of passes through the cycle. Hence the total radiation absorbed dose of solvent is limited and may be expressed as absorbed dose per pass times total number of passes (or recycling factor).

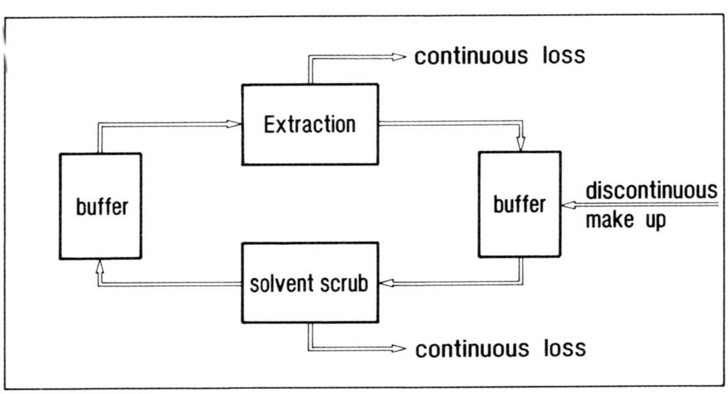

Fig. 10: The solvent cycle

During the above mentioned campaigns, a first cycle solvent inventory was taken daily to get an estimate of continuous solvent losses. On behalf of several reasons one will get moderate fluctuations of daily solvent inventory:

- the organic hold-up of extraction equipment will fluctuate,

- aqueous phase will be entrained in used solvent of the first extraction cycle and will accumulate in the used solvent buffer tank, from where it is sent to the solvent scrub waste from time to time,

- organic phase will be entrained in aqueous effluent of solvent scrubbers, it will be recycled to the used solvent buffer tank after the complete separation of phases.

Nonetheless, solvent losses of about 10 liters per day were determined, to be associated with a solvent recycling factor of 250.

To allow for an extrapolation to a commercial plant, the WAK solvent losses were traced down to different sinks. In doing so, one has to consider that WAK operates a diluent cycle in addition to its first solvent cycle. Used diluent from diluent scrubbing of aqueous uranium product and of aqueous raffinate of the second plutonium cycle (cf. fig. 9), respectively, is sent to the Karlsruhe Nuclear Research Center, which provides a facility to recover rather pure diluent. Since the recovery process is acidic, lauric acid type compounds supposedly will stay with the diluent. Recovered diluent will again be used for diluent scrubbing, thus closing the diluent cycle, or will be used for make-up of solvent stocks, linking the diluent cycle to the first solvent cycle. In addition used diluent from diluent scrubbing of aqueous raffinate of the first extraction is sent directly to the first cycle solvent (cf. fig. 9).

Taking into account all losses to aqueous streams either through entrained solvent or through dissolved TBP, as well as hydrolytic or radiolytic decomposition of TBP, the recycling factor of pure TBP in the WAK first solvent cycle is estimated to be 300, while the dodecane diluent recycling factor is only 200. Significant diluent losses are to be attributed to its rather high vapour pressure, however, radiolysis products with high molecular weight will not at all be lost to the vessel offgas.

5.4 TOTAL ABSORBED DOSE

From the radiation absorbed dose per pass of solvent through the first cycle (cf. table 7) and the evaluated recycling factor, the calculation of the total radiation absorbed dose is now straightforward (cf. table 8). A comparison of the numbers in the bottom line of table 8 shows a satisfactory agreement of the dose evaluated from analytically determined radiolysis products with the dose calculated from the solvent's irradiation history, validating the former method.

Campaign	Recycling Factor	D_α <-- (Wh/l)	$D_{\beta/\gamma}$ -->
6. MZFR	250	6	3
5. KWO	250	17	6
5. KWO	300	20	7.2
5. KWO Observed (cf. table 4)		28	1.2

Table 8: Total radiation absorbed dose of the first cycle solvent of WAK

Again we point out that about half of the α-radiation absorbed dose is due to the radiolysis of the solvent, entrained in the aqueous plutonium product of the first and second plutonium cycles (cf. section 5.2).

6 EXTRAPOLATED PERFORMANCE OF THE FIRST CYCLE SOLVENT OF A COMMERCIAL PLANT

As a major advantage of the proposed model of solvent radiolysis, one may infer the performance of the first cycle solvent of a future plant from already accessible data: the layout of the extraction equipment, the plant chemical flowsheet, and the isotopic composition of the fuel to be processed.

The first solvent cycle will comprise the first extraction cycle and the second and third plutonium purification cycle. The countercurrent extraction will almost exclusively be performed in pulsed column extractors.

Diluent washing will be provided for aqueous first, second, and third cycle plutonium products. Solvent entrained or dissolved in any other aqueous product or raffinate stream will be very effectively separated therefrom and recycled to the upstream decanter.

The expected dose rate of the plant feed solution and the expected specific dose rate of the fuel plutonium are already listed in tables 5 and 6, respectively. Next, the radiation absorbed dose per pass through the first solvent cycle is calculated as well as the total number of solvent passes through the cycle.

Following the method outlined in section 5.2, the absorbed dose per pass from the β/γ-radiation of the fission products will amount to

Uranium-FE: 0,014 Wh/l

which refers to the throughput of uranium fuel elements with a burn-up of 40 GWd/t and 7 years cooling time.

Likewise, α- (and β-) dose uptake through the radiation of extracted plutonium is evaluated. We arrive at the total absorbed dose per pass as listed in table 9.

Fuel (average burn-up)	D <--- (Wh/l)	$D_{\beta/\gamma}$ --->
Uranium FE (40 GWd/t)	0,054	0,016
+ 10 % MOX-FE (45 GWd/t)	0,090	0,018

Table 9: Radiation absorbed dose per pass of solvent through the first solvent cycle

Since the commercial plant will provide for diluent washing of aqueous plutonium product streams, the α-radiation damage to solvent through the entrainment path (cf. section 5.2) will be of minor importance.

The plant, however, will process uranium/plutonium mixed oxide fuel elements (MOX-FE) at a rate of 10 % of the total throughput per year.

In addition, the average burn-up of spent fuel may arrive at 45 GWd/t in the mid-nineties, hence α-radiation will play an even more dominant role.

The fission product content as well as the content of plutonium-241 of spent fuel will linearly increase with increasing burn-up, which will give rise to a linear increase of β/γ-radiation absorbed dose. However, the plutonium-238 content of spent fuel will increase with the burn-up squared (5). 10 % of MOX-throughput per year will increase the plutonium throughput by a factor of 1.2 with an additional factor of 1.1 for plutonium-238 content (9). Taking the appropriate factors into account, we arrive at a maximum absorbed dose per pass of solvent through the first cycle as listed in table 9.

Relative solvent losses from the first solvent cycle are expected to be even lower when compared to WAK. However, provisions made to recover and recycle entrained or dissolved solvent from aqueous streams will not recycle radiolysis products to the same extent. Thus, the recycling factor of solvent is estimated to be 300. A detailed analysis of loss patterns of pure TBP and of diluent, respectively, shows that there will be no significant differences between TBP and diluent recycling factors.

Given a spent uranium fuel specification of 40 GWd/t burn-up and 7 years cooling time, indefinite operation of the future plant would imply a maximum of 16 Wh/l α-radiation absorbed dose and about 5 Wh/l β/γ -radiation absorbed dose in the first cycle solvent (cf. table 10).

On increase of burn-up to 45 GWd/t and processing MOX-fuel elements at a rate of 10 % of the yearly throughput, the α-radiation absorbed dose would go up to 28 Wh/l, while the β/γ -radiation absorbed dose would effectively stay at 5 Wh/l.

Obviously, the first cycle solvent will suffer from radiation damage to the same degree as the WAK first cycle solvent already has. Hence the first cycle solvent will perform as well as the WAK solvent actually does.

Fuel (burn-up)	Recycling Factor	D_{α} <---	(Wh/l)	$D_{\beta/\gamma}$ --->
FCP				
Uranium FE (40 GWd/t)	300	16		4.8
+ 10 % MOX-FE (45 GWd/t)	300	27		5.4
WAK (cf. table 4)				
5. KWO calculated	300	20		7.2
5. KWO observed	-	28		1.2

Table 10: Maximum radiation absorbed dose of the first cycle solvent of a future commercial plant (FCP)

LITERATURE

(1) L. Stieglitz, R. Becker, R. Würtz:
"Die Bildung von brennbaren Gasen im PUREX-Prozess";
KfK-Nachrichten 3/82, p. 137

K.D. Kuhn, A. Sahm:
"Arbeitsschwerpunkte und Ergebnisse der DWK im gemeinsamen Entwicklungsprogramm";
4. Statusbericht des PWA am 05.11.1981;
KfK-Report No. 3290, p. 33

(2) L. Stieglitz, R. Becker:
"Chemical and Radiolytical Solvent Degradation in the PUREX-Process";
Nuclear Fuel Cycle, Vol. 2, p. 333
Verlag Chemie Weinheim, 1983

(3) A. van Schoor, M. Weishaupt, L. Stieglitz, R. Becker:
"Bildung von Dibutylphosphat im Lösungsmittel der WAK";
Jahrestagung Kerntechnik '86, Aachen
8.-10. April 1986, p. 257

(4) R. Becker, L. Stieglitz, H. Bautz:
"Development of Quality Tests for the PUREX Solvent";
Karlsruhe Int. Conf. on Analytical Chemistry in Nuclear Technology, June 3-6, 1985

(5) U. Fischer, H.W. Wiese:
"Improved and Consistent Determination of the Nuclear Inventory of Spent PWR-Fuel on the Basis of Time-Dependent Cell-Calculations with KORIGEN";
KfK-Report No. 3014, January 1983

(6) C. Michael Lederer, Virginia S. Shirley:
"Table of Isotopes, 7th edition";
John Wiley & Sons, Inc., New York, 1987

(7) G. Erdtmann, W. Soyka:
"Die γ-Linien der Radionuklide, Band 1";
Jul-1003-AC, Sept. 1973

(8) C.A. Blake, Jr.:
"Solvent Stability in Nuclear Fuel Processing - Evaluation of the Literature, Calculation of Radiation Dose and Effects of Iodine and Plutonium";
ORNL-4212, Oak Ridge National Laboratory (Nov. 1968)

(9) U. Fischer:
"Mehrfache Rückführung von Plutonium in thermischen Reaktoren";
Atomwirtschaft 31, p. 548 (1986)

The behavior of hydrazoic acid in PUREX process solutions under safety aspects

D. Ertel, H. Schmieder, A. H. Stollenwerk

SUMMARY

Several facts concerning the formation, properties and behavior of HN_3 in the PUREX process are discussed first in order to give background information referring to the safety aspects in question. The USAEC safety concept available since 1962 is briefly presented and subjected to a critical evaluation in the light of recent findings.

The recently performed systematic HN_3 analyses of process solutions from the Karlsruhe reprocessing plant (WAK) have demonstrated that most of the HN_3 occurs in organic solutions whose HN_3 vapor pressure is markedly lower than that of aqueous process solutions, and that critical HN_3 concentrations do not occur anywhere. Moreover, rates of HN_3 formation were determined for the first time in the 1B and 2B contactors (mixer settlers) of WAK. Therefrom, the conclusion can be drawn that in the conventional mode of operation using external U(IV) feed much more HN_3 is formed than in advanced electrochemical process engineering with insitu production of U(IV).

Summarizing it can be stated that the US safety concept, despite the extremely conservative limits it sets, does not impose any restrictions whatsoever on practival process engineering and, consequently, guarantees a very high level of safety.

The subjects to be treated in further planned experiments on the behavior of HN_3 in the PUREX process will be briefly outlined.

Dr. D. Ertel, Dr. H. Schmieder, Institut für Heisse Chemie, Kernforschungszentrum Karlsruhe, Postfach 36 40, D-7500 Karlsruhe 1, Federal Republic of Germany
Dr. H. A. Stollenwerk, Wiederaufarbeitungsanlage Karlsruhe Betriebsgesellschaft mbH, Postfach 2 20, 7514 Leopoldshafen-Egg. 2, Federal Republic of Germany

1 INTRODUCTION

The most conspicuous property of hydrazoic acid (HN_3) and of some of its salts (azides) is the explosiveness.

Although this property is realized only under well defined physical conditions and therefore should not be considered as being of general validity, it must be taken into account in all cases where particularly exacting standards have to be applied to safety.

Since in the PUREX process hydrazoic acid may be formed as an oxidation product of hydrazine, it must be examined whether a safety risk is involved. For this, the process conditions have to be compared with the limits of HN_3 explosiveness and it must be reliably cleared up whether or not there are overlappings. Responsible experts did so as early as 25 years ago when the U-IV/hydrazine concept was introduced into PUREX based process engineering /1,2/. After extensive safety analyses they stated a negative result: the probability of an HN_3 explosion occuring in the process was considered very low. Besides, the authors at the time fixed safety limits below which explosions of HN_3 can be completely ruled out.

Despite practical experience meanwhile accumulated according to which no HN_3 explosions have accurred in reprocessing plants /3/, the modern tendency to reevaluating ancient safety standards has also shed light on the so-called HN_3 problem in the PUREX process. Notwithstanding the fact that the situation has essentially remained unchanged some critics thought to have discovered gaps in argumentation, for instance the lack of knowledge of the HN_3 concentrations in the various process streams.

In this situation we have taken efforts for several years to elucidate facts in an objective manner, either by presentation and interpretation of already established knowledge or by specific experimental studies on real PUREX process solutions. The status of these efforts will be reported below.

2 FORMATION OF HN_3

In the Purex process hydrazoic acid is formed during oxidation of hydrazine with nitrous acid, HNO_2, or with nitrate:

$$N_2H_5^+ + HNO_2 \longrightarrow HN_3 + 2\ H_2O + H^+ \qquad (2.1a)$$

$$HN_3 + HNO_2 \longrightarrow N_2O + N_2 + H_2O \qquad (2.1b)$$

$$2\ N_2H_5^+ + NO_3^- \longrightarrow HN_3 + 3\ H_2O + N_2H_3^+ \qquad (2.2a)$$

$$HN_3 + N_2H_3^+ \longrightarrow NH_4^+ + 2\ N_2 \qquad (2.2b)$$

The status of HN_3 as an intermediate product is typical. On account of the competitive nature of the reactions of formation and decomposition the HN_3 yield - dependent on the actual conditions - passes a maximum which, however, attains only a fraction of the stochiometrically possible yield.

The reactions (2.1a) and (2.1b) describe the classical situation of hydrazine oxidation with HNO_2: Hydrazine is present in excess (holding reductant, nitrite scavenger), and HN_3 formation is a function of the HNO_2 source strength.

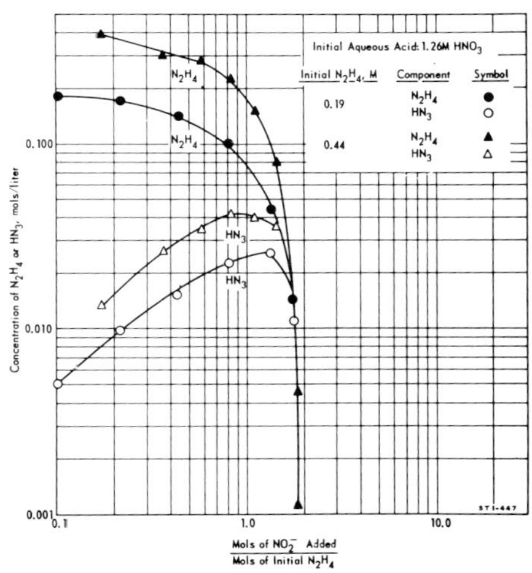

Fig. 1: Destruction of nitrite in the absence of the organic phase

Figure 1 illustrates the course of the reaction by the example of the stepwise reaction of a given amount of hydrazine with nitrous acid /2/. It can be seen that at first HN_3 is formed preferably, then reaches a maximum value and, finally, is destroyed again by HNO_2 in excess. The maximum yield occurs at the molar ratio N_2H_4/HNO_2 = 1 and amounts to about 12% of the initial hydrazine concentration. Thus, the given N_2H_4 concentration sets a limit on the maximum HN_3 concentration possible, whereas the reacted amount of HNO_2 determines the HN_3 concentration in the process. It should be added that in the real PUREX process a substantial fraction of HN_3 is extracted into the TBP/alkane phase.

The other reaction of formation (2.2a) is the oxidation of hydrazine by nitrate in presence of technetium as a catalyst. Although technetium has always been present in the PUREX process solutions as an (extractable) principal fission product, this reaction has been studied in more detail just recently: it has been found to be rather complex /4,5,15/. The total reaction according to the formula given here shows that the primary oxidation products HN_3 and $N_2H_3^+$ react further to give ammonium and nitrogen.

The reactions leading to the destruction of hydrazoic acid are essentially unknown as are the exact formulae of the technetium intermediates; however, the nature of the final products (NH_4^+, N_2) has been demonstrated /4/.

3 PROPERTIES OF HN_3

The HN_3 properties which are relevant to the PUREX process have been compiled in Tab.1:

Acidity exponent	pK_a 4.7
Boiling point	Kp_{760} 35.7° C
Vapor pressure at 24.7° C	p 509 mm Hg
Density	D_4^{25} 1.09 g/cm^3
Solubility in 30% TBP/alkane	$D_{o/a}$ 2 - 10
Explosion limit in aqueous solution	ca. 17 w/o

Table 1: Selected properties of HN_3

Hydrazoic acid is a relatively weak acid of the approximative strength of acetic acid. The chemical behavior of its salts (azides) sometimes resembles that of hydrochloric acid. A characteristic is e.g. the formation of scarcely soluble azides with Ag^+, Hg_2^{2+}, Pb^{2+} and Tl^+. HN_3 has a remarkably low boiling point which means that at room temperature it is approximately as volatile as diethylether. This property has a considerable impact on the HN_3 safety concept (see below).

A very noticeable experimental result, which is also of considerable importance to safety, was communicated by one of us (D.E.) at the Obertraun Seminar 1984 /6/: The HN_3 vapor pressure of organic process solutions is much lower than that of correspondent aqueous solutions which is probably the result of $HN_3 \cdot TBP$ complexing /10/.

Special attention should also be paid to the good solubility of HN_3 in tributyl phosphate/alkane mixtures; it depends on the U, Pu loading of the extractant /2/. The solubility in TBP of HN_3 is essential for its distribution and behavior in the PUREX process (also HNO_2 is well soluble in TBP, but not hydrazine).

The molecular structures of HN_3 and its salts are exposed to high stress.

Whereas the HN_3 molecule is characterized by a linear grouping of the N atoms with different bond lengths (1.13 A for the N-N bond and 1.24 A for the N-NH bond) and can be described by the following resonance formulae:

the azide anion is linear and symmetric in its structure

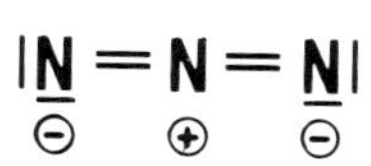

and has identical N-N lengths of 1.18 A.

The bonding character between anion and cation is the actual reason for the tendency to explode: Whereas the rather covalent heavy metal

azides (including HN_3 itself) might undergo violent explosions, the salt type alkali azides, for example, are completely harmless. During the explosive decomposition of HN_3 into nitrogen and hydrogen large amounts of energy are set free:

$$2\ HN_{3(l)} \longrightarrow 3\ N_2 + H_2 + 526\ kJ$$

The lower limit of explosion of aqueous HN_3 solutions is frequently indicated to occur at about 17 weight percent (4.7 mole/l) /2/.

However, it appears from the original results compiled in Tab.2 that at about 17% HN_3 content one can speak at best of a combustion or deflagration because the rate of propagation is only 0.14 m/s. By definition, the rates of propagation of explosions range from 1-1000 m/s; at still higher rate the reaction is called a detonation.

Concentration of HN_3 g/100 g solution	Normality of solution mole/l	Velocity m/s
80	22.1	8.500
70	18.0	7.300
67	17.2	7.100
65	16.8	1.700
61	15.7	630
60	15.4	384
54	13.8	233
48	12.2	40
38	9.8	2.5
30	7.3	1.0
16.7	4.0	0.14

Table 2: Velocity of propagation of explosion in hydrazoic acid solutions
(taken from Ref./7/)

Strictly speaking, HN_3 explodes only above 30% concentration (7.3 mole/l) and undergoes detonation beginning at about 62% (16 mole/l).

(If we continue to speak of the lower limit of explosion occuring at 17%, this is to underline the conservative nature of the safety concept derived from it).

4 BEHAVIOR OF HN_3 IN THE PUREX PROCESS

HN_3 is formed in the hydrazine carrying separation stages of the process. On account of its good solubility in TBP it is spread downstream up to the backextraction stages and the solvent wash, respectively. Above all the organic process streams BU and CW transport hydrazoic acid whereas the corresponding aqueous process streams BP and CU contain but relatively low amounts of HN_3 (cf. Tab.3).

The extraction and backextraction of U(VI) and Pu(IV) are not influenced by the presence of HN_3 /8/, although there are indications of the existence of soluble U and Pu azido complexes resulting from spectrophotometric investigations /9/. However, the respective compounds are formed only at molar ratios $>$ 1 of HN_3/U and HN_3/Pu, respectively, i.e. under conditions not encountered in the PUREX process.

The formation of solid U or Pu azides, e.g. in the alkaline solutions of the solvent scrub, can be safely excluded because, from a thermodynamic point of view, under these conditions in the fluid carbonato complexes, diuranates or hydroxides are preferably formed /2/. Regarding the formation of alkali azides in the solvent wash it is further worth mentioning that also here the HN_3 vapor pressure is drastically reduced.

5 HN_3 SAFETY CONCEPT

Recommendations on the explosion proofness of hydrazoic acid in the PUREX process were publicized in the USA as early as in 1962 /1,2/. They are based on the just discussed "lower explosion limit" of 4.7 mole/l aqueous HN_3.

In the USA a general concentration limit of 0.05 mole/l HN_3 is recommended for PUREX process solutions. This concentration which is surprisingly low at first glance results from the high HN_3 vapor pressure because the equilibrium gas phase overlying an aqueous

0.05 mole/l HN_3 solution contains so much HN_3 at 20° C that condensation may cause an HN_3 concentration of 4.7 mole/l. This argument does not hold for solutions at 40° C - which is nearer to reality - whose H_2O partial pressure is so high that critical concentrations of HN_3 in the condensate are not attained.

A further recommendation concerns the HN_3 concentration in the gas phase which should be < 2.5 volume percent.
Such a limit concentration has an energy potential which would be just sufficient to heat the gas mixture to 290° C (adiabatic); at this temperature thermal dissociation of HN_3 starts.
A gas phase containing 2.5 volume percent HN_3 is in equilibrium with an 0.2 mole/l HN_3 solution. Since such a concentration of the solution is higher than the 0.05 mole/l limit, it is "interdicted" on safety grounds and does not occur in the PUREX process.

After these rather theoretical considerations let us turn again to the real PUREX process. Usually, hydrazine concentrations of a few tenths of mole/l are used in the separation stages. If 0.4 mole/l is assumed, a 12% reaction yield (cf. Section 2) would give a maximum of 0.05 mole/l hydrazoic acid. In other words: The HN_3 concentrations to be expected a priori in the PUREX process are strictly limited by the mere boundary conditions applicable to this process and, therefore, should hardly attain the actually very conservative safety limit, let alone exceed it.

6 HN_3 CONCENTRATIONS IN THE REAL PUREX PROCESS

So far only few data have been known on the HN_3 concentrations in PUREX process solutions.

A value of 0.012 mole/l formerly indicated in the USA /1/ relates to a PUREX variant used in Np/Pu separation and therefore better provides an idea of the general level of concentration than a concrete information does.

In recent years we have developed an efficient analytical method /11/.

The HN_3 assays which have become feasible by application of this method to the process solutions of various R+D facilities of the KfK Hot Chemistry Institute yielded a maximum of 0.039 mole/l HN_3 in the BU stream of the 1B mixer settler of the MILLI facility /12/.

Recently, comprehensive systematic analyses of process solutions from the WAK reprocessing plant have been made for the first time /13/. The objective of these investigations has been the determination of the HN_3 contents of all relevant process streams under real plant conditions as well as the dtermination of HN_3 source terms. A special feature of the WAK flowsheet (Fig.2) consists in linking the 1B contactor to the 2B contactor whose extractants are passed through a joint solvent scrubber. Furthermore, it should be noted that the 1B mixer settler is operated with conventional external U(IV) feed whereas the 2B tank (mixer settler) works electrolytically. (Since November 1986 the WAK-1B-bank is operated by electrolysis, too).

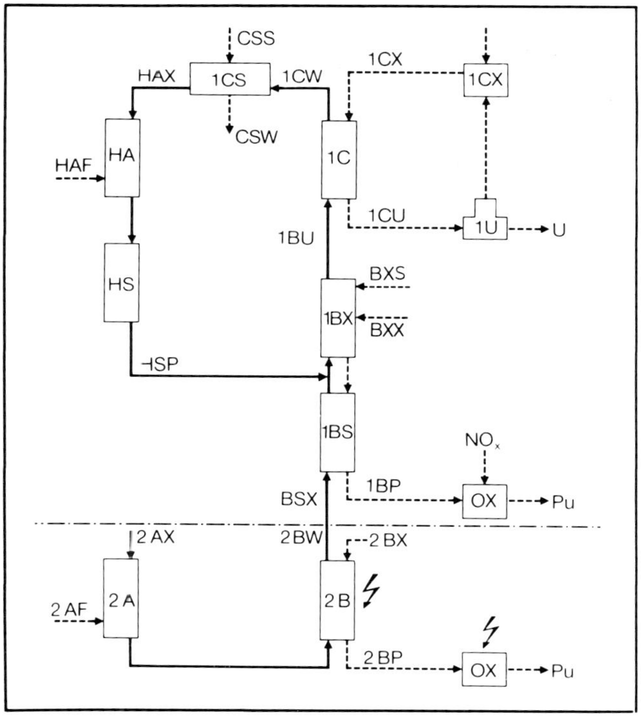

Fig. 2 WAK-1st cycle flowsheet and HN_3 sampling points

Table 3 shows the results of analysis. It can be seen at a glance that the HN_3 safety concentration of 50 mmole/l is not attained at any point and even clearly underrated everywhere. Except for 2BW, the HN_3 contents of all other process streams attain 10 ± 5 mmole HN_3/l. The

HN_3 feed of 9 mmole/l with the external U(IV) feed (BXX) is remarkable.

Flow	HN_3 concentration (mmole/l)		Flow rate (l/h)	HN_3 mass flow (mmole/h)
BXX	9.0	(5)[b]	10	90
1BP	5.8	(3)	16	93
1CU	5.8	(7)	100	580
1CX	5.0	(7)	100	500
CSW	15	(4)	20	300
1BU	11.8	(7)	100	1180
1CW	9.2	(9)	100	920
2BW	27.7	(6)	9	249

Table 3: HN_3 concentrations and mass flow in WAK solutions[a)]
a) reprocessing of high-burned power reactor fuel; initial hydrazine concentration in BXX, BXS and 2 BX 0.4 mole/l
b) number of samples analyzed

The mass flows are even more interesting. The main carrier fluid is the 1BU stream with > 1000 mmole/h whereas in the corresponding 1BP stream only just under 10% of the 1B flow is attained.

The high concentration in 2BW produces but a relatively low mass flow. It is interesting to note that in the U evaporator (inlet 1CU, outlet 1CX) only little HN_3 is lost; we had rather expected high losses due to oxidation and evaporation, respectively.

Obviously, the solvent scrubber did not work properly during the sampling campaigns of several days duration because only about one third of the amount present was scrubbed (1CW/CSW).

If one strikes the balance of HN_3 inputs and outputs of the two separation contactors, the following rates of formation are obtained:

1B - mixer settler (conventional U-IV): 934 mmole/h

2B - mixer settler (electrochemical): 270 mmole/h

These remarkably differing results can be evaluated with difficulties only because the conditions in terms of equipment and process chemistry are dissimilar and, in this respect, the results cannot be generalized. But still, the different U(IV) reduction potential (1B > 2B) seems to play an essential role in HN_3 formation. On this assumption it can be concluded that during conventional operation with external U(IV) feed much more HN_3 is formed than by advanced electrochemical process engineering.

7 CONCLUSIONS AND PROSPECTS

The results of the experimental investigations presented here offer for the first time reliable information about real HN_3 concentrations occuring in various PUREX process streams, HN_3 rates of formation under typical plant conditions, and the greatly differing HN_3 vapor pressures of correspondent organic and aqueous process solutions. These results prove that some considerations underlying the US safety concept have to be considered as being extremely conservative. If, nevertheless, no restrictions at all result for practical process engineering, this is a proof of the high degree of safety offered by this concept. This opinion is also confirmed by experience accumulated in practical work where no HN_3 accidents have occured in reprocessing plants /3/.

By proving that in the streams of the PUREX process critical HN_3 concentrations do not exist we have, for the time being, attained our main goal.

But still questions remain to be answered which will be dealt with in further experimental investigations:

1. Precise measurements of HN_3 vapor pressures of organic and aqueous process solutions

   It is proposed to confirm in quantitative terms the but qualitatively established fact that there are great differences in the HN_3 vapor pressures.

2. Determination of the reaction rates of HN_3 and HNO_2 in TBP/alkane. At present, it is only known that this reaction proceeds at a much slower rate than in aqueous solution.

3. HN_3 in the vessel off-gas system /14/

Due to the HN_3 vapor pressure of aqueous process solutions some of this inventory reaches the vessel off-gas. It has to be examined whether HN_3 in the vessel off-gas system of a reprocessing plant can be measured at all. As a matter of fact, it is probably destroyed at once due to the always simultaneous presence of NO_x and H_2O in the vessel off-gas system.

$$(NO + NO_2 + H_2O \longrightarrow 2\ HNO_2;$$
$$HN_3 + HNO_2 \longrightarrow N_2 + N_2O + H_2O).$$

4. Investigation of the reactions taking place in the gas phase system $HN_3/NO_x/H_2O$

It is proposed to clarify by model investigations whether HN_3 can exist at all under these conditions.

ACKNOWLEDGEMENT

We are indebted to and would like to appreciate the contributions of our staff members

I. Souba
K. Weiß
S. Lang
J. Scherwitzl
G. Horn

who with a lot of engagement and intuition have developed the method of HN_3 analysis and advanced it to perfection and who have performed a great number of analytical assays.

LITERATURE

(1) E.K. Dukes, R.M. Wallace, USAEC-Report DP-728 (1962)

(2) C.S. Schlea, M.R. Caverly, H.E. Henry, W.J. Jenkins, USAEC-Report DP-808 (1963)

(3) W.D. Bond, in "Light Water Reactor Nuclear Fuel Cycle" CRC Press, Inc., Boca Raton/Fa. (1981)

(4) J. Garraway, P.T. Wilson; J. Less-Common Metal 97 (1984) 191

(5) V.S. Koltunov, V.I. Marchenko et al., Atomnayà Energiya 60 (1) (1986), 35

(6) D. Ertel, Winter-Seminar "On new developments in the Field of Reprocessing", Obertraun/Austria, March 4-10, 1984

(7) J. Taylor "Detonation in Condensed Explosives", Clarendon Press Oxford, 1952, p.159

(8) P.C. Mayankutty, N.S. Pillai, S.S. Shinde, S. Ravi, M.N. Nackarni, Symp. on Solvent Extraction of Metals, Bombay/India, Febr.1-2, 1979

(9) F.G. Sherif, A.M. Awad; J. Inorg. Nucl. Chem. 19 (1961) 94 und 24 (1962) 179

(10) A.D. Kelmers, D.N. Browning; Inorg. Nucl. Chem. Letters 13 (1977) 505

(11) D. Ertel, Atomkernenergie/Kerntechnik 47 (1985) 87

(12) D. Ertel, G. Horn, K. Weiss; Internat. Conf. on Nuclear and Radiochemistry, Lindau/Germ., Oct. 1984

(13) D. Ertel, M. Heilgeist, A.H. Stollenwerk, H. Weishaupt, Internat. Conf. on Nuclear and Radiochemistry, Beijing/China, Sep. 1-5, 1986

(14) H. Schmieder, D. Ertel, unpublished results, 1986/87

(15) Work performed by B. Kanellakopulos and coworkers at the Institute of Hot Chemistry, KfK: C.P. König, PhD Thesis, Univ. of Heidelberg, 1983; N. Boukis, PhD Thesis, Univ. of Heidelberg, 1985.

New procedures to verify the input analysis of nuclear fuel reprocessing

L. Koch

SUMMARY

During the last 15 years new techniques have been developed to determine the input and output of fissile material of a nuclear fuel reprocessing plant. Isotope correlation technique has become a versatile diagnostic tool, automation of mass-spectrometric isotope dilution analysis speeds up input verification as does K-edge densitometry for the output. The latter technique when combined with X-ray fluorescence analysis could be applied more generally.

Dr. L. Koch, Commission of the European Communities, Joint Research Centre, Karlsruhe Establishment, European Institute for Transuranium Elements, Postfach 23 40, 7500 Karlsruhe, Federal Republic of Germany

1 INTRODUCTION

In its broader sense, "safety" includes the detection of diverted fissile material. In the Federal Republic of Germany this so-called "Nuclear Material Safeguards" is a task of EURATOM. The ECSAM laboratory (European Safeguards Analytical Measurements) of the European Institute for Transuranium Elements has been charged with verifying the fissile material content in samples taken by EURATOM inspectors.

For this purpose new techniques and instruments were developed in the seventies, which in particular in this laboratory were sponsored to a large extent by the project "Spaltstoffflußkontrolle" of the Kernforschungszentrum Karlsruhe (KfK). The new procedures matured during this cooperation are:

- isotope correlation technique combined with gravimetric input analysis
- automatic mass-spectrometric isotope dilution analysis
- K-edge densitometry

2 ISOTOPE CORRELATION TECHNIQUE

The formation and transmutation of actinides and fission product nuclides in a spent nuclear fuel are inherently connected, so that from the abundance of a selected monitor nuclide the corresponding parameters of the fuel can be obtained: for fissile material control the concentration of the fissile material. Earlier studies proved that the isotopic composition of the fission gas krypton correlates well with the fission source ratio i.e. the consumption of Pu-239/U-235 in a spent fuel /Fig.1/.

A simple analysis of the off-gas of the reprocessing plant with a measurement of the isotopic abundance of the fission gas nuclides would allow a determination of fuel parameters such as the fission source, the burn-up, Pu build-up, U-235 depletion etc. The same correlations of course are observed with the actinide isotope abundances. The verification of a reprocessing input analysis produces these data and therefore their use for isotope correlation purposes does not require additional analyses. An international experiment relying on the reprocessing input data obtained by four different laboratories was used to check the potential of the isotope

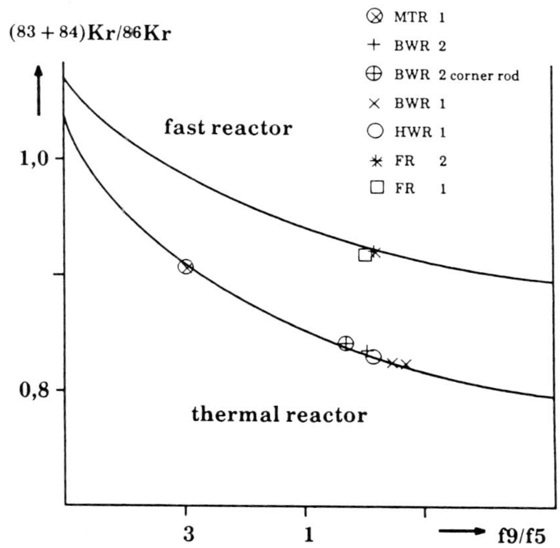

Fig.1: ^{239}Pu and ^{235}U fission source ratio (f9/f5) plotted against $^{(83+84)}Kr/^{86}Kr$ ratio

Method	Pu/U	
Isotope correlations:		
X = D-U-235	± 1.0 (1)	+ 1.0
Pu-242/240	± 1.6	+ 0.3
Cm-244 IMA	± 1.1	+ 0.9
Xe-132/131	± 2.0 (2)	+ 0.4
Cs-134/137	± 1.7	+ 1.6
100 - Pu-239 (a/o)	± 1.7	+ 0.7
Pu-(242-239) / (240-240)	± 1.6	- 0.3
ANOVA	± 1.3	---

Tab.1: Mean variation coefficient(%) and bias(%) of nuclide concentrations as evaluated by isotope correlations

correlation technique /1/. The correlations used in addition to those obtained from the isotope analysis of uranium and plutonium are listed in tab.1. The accuracy was 1-2% which at that time corresponded to the overall accuracy of the analytical techniques. Later, a similar exercise conducted at the reprocessing plant at La Hague obtained higher accuracies /2/. By opposition to simple verification measurements the isotope correlation technique is able to determine and explain the source and cause of an error. To demonstrate this capability concealed modifications of a set of 29 data were made during a second experiment at La Hague. The isotope correlation technique developed in this laboratory /3/ was able to detect all the modifications and could determine the cause of the error in 10 of the 11 cases.

The isotope correlation technique cannot directly determine the masses of uranium and plutonium in a plant but has to rely on additional information e.g. on the initial fuel weight. The technique will therefore not be accepted as a primary verification tool, but rather be used to check the consistency of verification measurements and to detect and explain possible errors.

3 AUTOMATIC LABORATORY

The present automatic laboratory has evolved over the last fifteen years. The automation followed three objectives:

- automatic recording, reduction and transfer of data,
- automatic operation of instruments generating safeguards relevant data,
- automatic devices to handle and condition samples for the subsequent measurement.

At present the data processing is done by a network of mini-computers which is connected via a transfer computer to a central computer /Fig.2/. The variety of computers used is explained by the fact that presently instruments are sold together with mini-computers with software which cannot easily be adapted to other computers. Impending improvements of the central computer will result in further changes in the local area network.

The advantages of automatic data processing are manifold. During the analysis of reprocessing input, data are generated by different instruments (balance, alpha-spectrometer, mass-spectrometer) which have not only to be recorded but reduced (e.g. alpha- and mass-spectra) before they can be evaluated. Software developed for this specific task collects the data of a single analysis from the different computers into a main databank. When the reprocessing input analysis is initiated the scope of information to be collected is defined automatically. As soon as expected information arrives the evaluation is started. The procedure is speedy, avoids

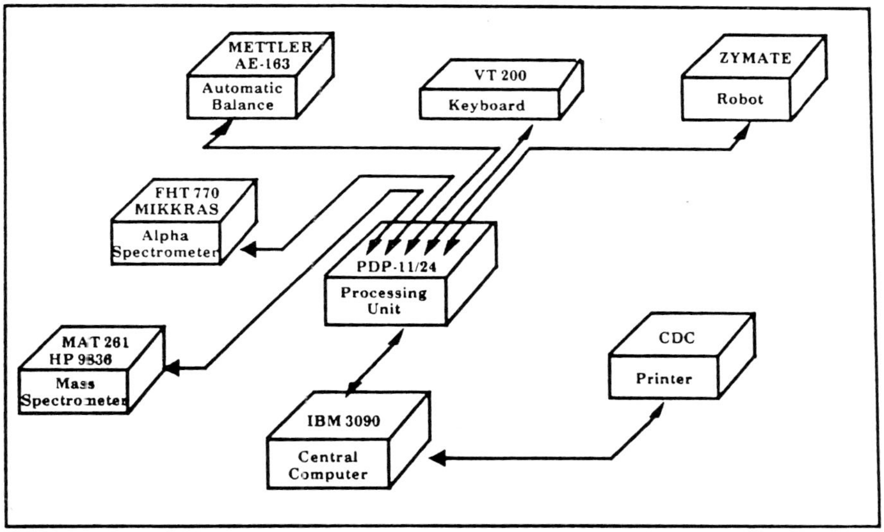

Fig.2: Scheme of Automatic Laboratory

transcription errors (as observed in a recent off-line evaluation) and tamper resistant - a noteworthy aspect relevant to a safeguards analysis.

When the automation started several years ago there were no automatic instruments for our purpose on the market. Therefore we developed the first fully automatic mass-spectrometer /4/, an alpha-spectrometer with an automatic sample changer and data recording as well as an automatic balance. In the meantime improved instruments are commercially available and consequently have been implemented in our laboratory.

The balances used are Mettler type AE163. Both are operated by foot-switches which avoid unnecessary movements of the gloves (of the glove-boxes) which could cause air turbulences and would interfere with the weighing process. Two alpha-spectrometers Frieseke & Höpfner, type Mikras with sample changers type FHT 770m are in use. The automatic sample changers hold 60 samples. The alpha-spectra are stored in a file where they can be visually inspected on screen before evaluation. Two mass-spectrometers of the type Finigan MAT 260 and 261 analyse sequentially 13 samples each. All steps from pre-heating the two rhenium filaments to the final evaluation are computer controlled. The MAT 261 is equipped with nine Faraday-cup detectors by which all masses of U or Pu can be recorded simultaneously. The interference of isobars (of ^{241}Pu-Am and ^{238}Pu-U) is monitored by the corresponding isotopes ^{235}U, ^{239}Pu, ^{243}Am.

In order to speed up the sample conditioning and to improve the reproducibility of the process two further automats are in use: an automat to separate U and Pu by ion-exchange and an analytical robot to aliquot sample and spikes, to mix the isotopes and to condition the solution for the subsequent separation.

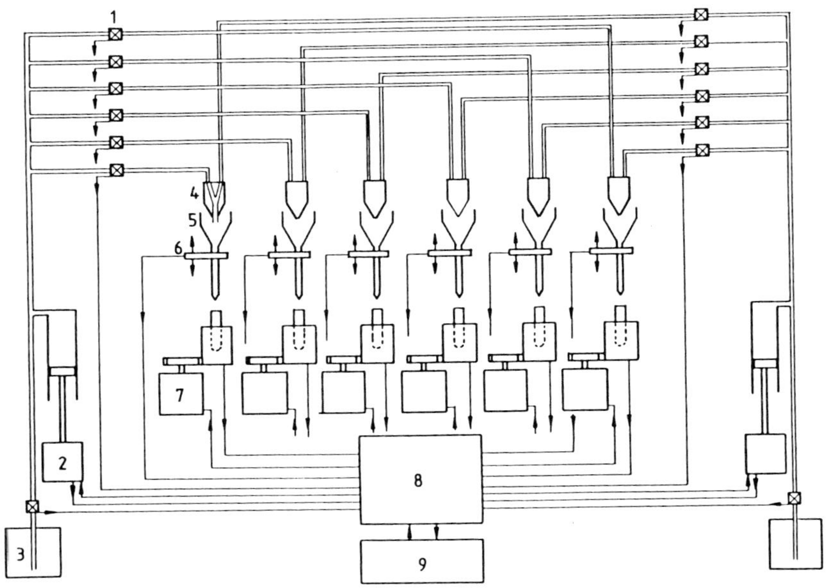

Fig. 3: Scheme of automatic conditioning of U and Pu. Valve (1), Pump (2), Reagent Flask (3), Injector (4), Column (5), Photocell (6), Sample Changer (7), Interface (8), HP Computer (9)

The separation of U and Pu by ion-exchange follows a standard procedure /5/. In the device - a prototype developed in our institute - up to 6 anion-exchange columns are separately controlled by a computer /Fig.3/. The eluate and wash solutions are dispensed to the individual columns at a rate governed by the flow rate of the column. Separate fractions for U and Pu are collected which are used to prepare the alpha- and mass-spectrometric samples /6/.

An analytical robot (Zymate) has been adapted to glove-box work /Fig.4/. The major periphery devices comprise: balance, mixer (for liquids), liquid dispenser and heating plate. The layout shows the partitioning of the box into three compartments: one used for the introduction of potentially contaminated samples, the main operation area and a hood for working in a corrosive atmosphere. The robot processes two exchangeable hands, one to open and handle the possibly contaminated sample bottles and the second for the subsequent work. To avoid corrosion the robot is flushed internally with argon. Since the conditioning of the samples is plant- and fuel-dependent several standard programmes are used, which allow e.g. for different methods of ensuring isotope exchange based on treatments with NH_4F, H_2O_2,

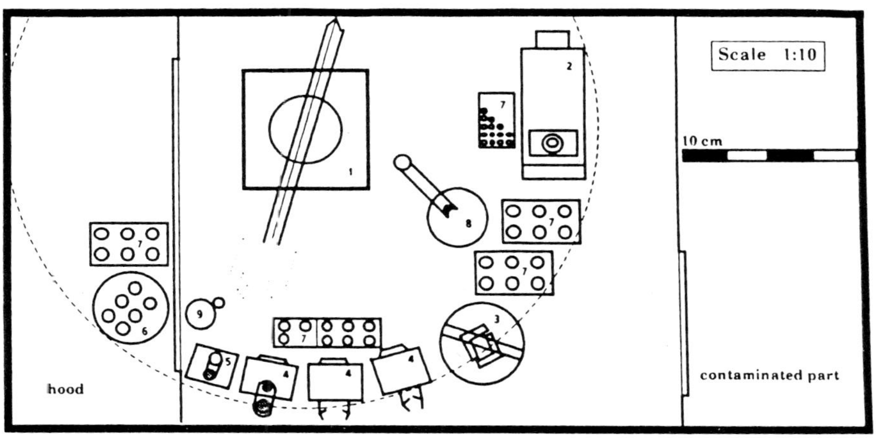

Fig. 4: Layout of an analytical robot (1) in a glove-box, balance (2), decapper (3), exchangeable hands (4), mixer (5), heating plate (6), racks (7), waste (pipettes, etc.) (8), liquid dispenser (9)

The automation of the input analysis has had - of course - an impact on the quantity and quality of personnel and analysis.

If an instrument is automated its capacity will increase by at least a factor of 2 if it can work unattended over night. We use this drastic increase in the capacity to broaden the quality control and to cope with peak-loads, which can be up to 150% of the base load. An increase in the analytical capability is also needed for the repeat measurements following more stringent error limits of the results which have been imposed recently.

Presently the automatic laboratory is set up so that it can handle three duplicate samples at once. Hence 6 solutions have to be separated by the automatic Pu/U-separation device which thus produces 12 samples for the mass-spectrometry. For this reason the turret of the mass-spectrometer takes 12 samples plus 1 isotopic standard for a single loading. It should be noted that the use of ^{244}Pu spike makes the analysis of an unspiked sample unnecessary (an unspiked sample had to be measured previously when employing ^{242}Pu).

The operation of the automats requires somewhat different personnel qualifications. All routine work (usually requiring lower skills) are taken over by the automats. The operator should possess basic knowledge in informatics and - if possible - electronics. As a result of the automation less qualified personnel had to be trained to receive the necessary skills. We have found that the automation has led only to a marginal decrease in the number of personnel and at the same time to a drastic increase in the sample throughput. This has enabled us to meet the increased demand for verification analyses (we receive samples from several reprocessing and fuel fabrication plants).

Since the reproducibility to which an automat works is much better than that usually obtained by human work the precision of an analysis is enhanced. The higher analytical capacity allows for more analysis repetitions. The accuracy of the analysis is ensured by applying quality controls for the error sources of the instruments (e.g. by measuring isotopic standards) and measures the composite errors by introducing secondary standards into the analytical scheme. At present the ESARDA target values /7/ of 0.5 % relative standard deviation for Pu and U assays can be easily met.

4 K-EDGE DENSITOMETRY

K-edge densitometry is a versatile non-destructive technique to measure the concentration of uranium and plutonium in solution. The device developed at KfK /8 + 9/ analyses the uranium and plutonium concentration in the output and the uranium concentration in the input of a reprocessing plant. When combined with X-ray flourescence this technique is able to determine low concentrations of plutonium, too, as it is the case in the reprocessing dissolver solution. The two instruments have been installed at the European Institute for Transuranium Elements for a thorough field testing.

K-edge densitometry is based on the abrupt change of the transmitted X-ray intensity at the absorption-edge of the atomic K-shell. The step-like transition is seen from the K-edge spectra taken for different actinide elements /Fig.5/.

The fractional density, $_{A}$ of the dissolved transuranium element relates to the measured transmission ratio, R (determined directly above and below the absorption-edge energy) as follows:

$$\rho_A = \frac{\ln R}{\Delta\mu_A \cdot D} + \frac{\Delta\mu_M}{\Delta\mu_A} \cdot \rho_M$$

where $\Delta\mu_A$ and $\Delta\mu_M$ are the differences in the photon mass attenuation coefficients (at the two transmission energies) for the transuranium element and the matrix. ρ_M is the matrix density and D the sample thickness.

4.1 K-EDGE DENSITOMETRY FOR OUTPUT VERIFICATION OF REPROCESSING

The instrument uses an X-ray tube as a photon source, which operates at voltages of about 140 to 150 keV to produce photons in the required energy range of about 110 to 125 keV. The X-ray beam has to be suitably filtered and collimated. The concentration of the dissolved transuranium sample is measured in a disposable glass-cell of which the path-length is known to ± 2 um. The sample-cell together with the collimator are contained in a glove-box. A high resolution, small

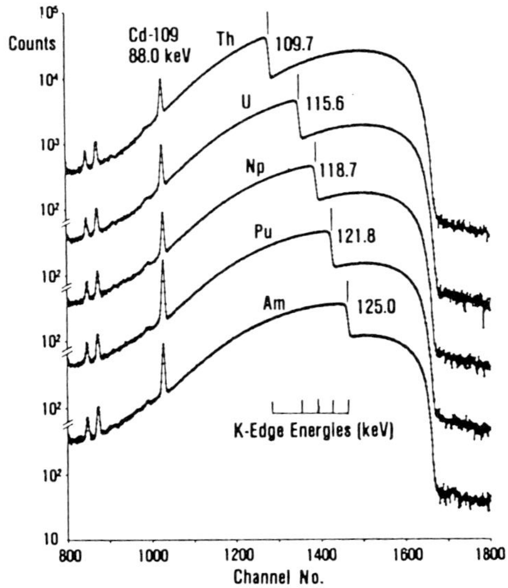

Fig. 5: K-edge spectra of actinide elements. The gamma line at 88.0 keV originates from ^{109}Cd source mounted close to the detector. It serves as a reference line for digital stabilization of the analogue electronics

planar Ge-detector is used to measure the photon transmission spectra, which are collected and analysed by a computerised multichannel-analyser with standard peripheries.

The procedure to determine routinely the element concentrations of uranium or plutonium is rather simple:

- the liquid sample has to be brought into a disposable glass-cell
- the glass-cell is placed into the beam-path at a defined position
- the absorption spectra are recorded and evaluated automatically.

The stability of the K-edge densitometer is routinely checked by a standard solution containing 220 mg U/ml in a sealed cell. For 45 checks over a period of 20 month the variation was 0.16%. For the case of plutonium the standard solution had a Pu-concentration of 178 mg/ml. Here the observed stability was 0.24% for 529 checks over a period of 38 months.

During a period of about 1 year output samples from the WAK reprocessing plant have been assayed for their plutonium and uranium concentration, and for Pu isotopic composition, in several consecutive batches. For the Pu concentration analysis we observed a bias of +0.01% with a relative standard deviation of 0.38% for 23 analyses (1 outlier). The corresponding errors for the uranium determination were -0.5 ± 0.15%. It should be noted, however, that the stability checks showed a variation of only 0.16%, as given above.

4.2 K-EDGE/K-XRF X-RAY DENSITOMETER FOR INPUT ANALYSIS

The concentration of uranium in the solution of a dissolved spent fuel can be analysed without interference from its radioactivity. In fig.6 the passive spectrum of the typical LWR fuel is given compared with the absorption edge spectrum.

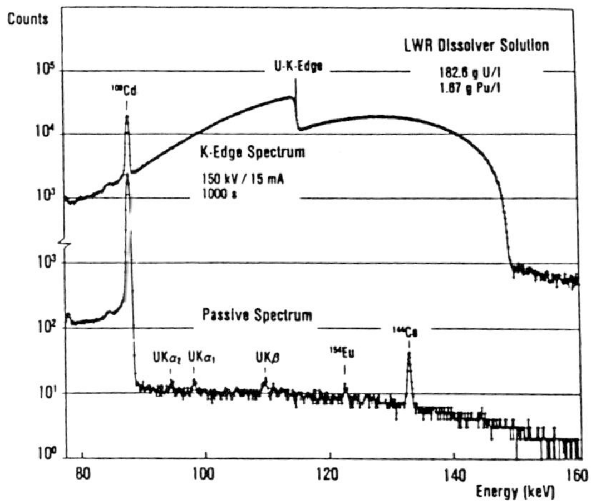

Fig. 6: Absorption edge spectrum (top) and passive spectrum (bottom) from a representative LWR dissolver solution (fuel burnup 35 GWd/tU, cooling time 3 years)

The plutonium concentration is in the order of 1% of uranium so that X-ray fluorescence (XRF) is used instead. This well-proven technique is used in the energy dispersive mode. A spectrum for a typical spent LWR is given in fig.7. The combination of the two techniques eleminates the need of an internal standard usually required for X-ray fluorescence analysis. Hence the accuracy of uranium and plutonium analysis is mainly determined by the capability of the XRF-technique.

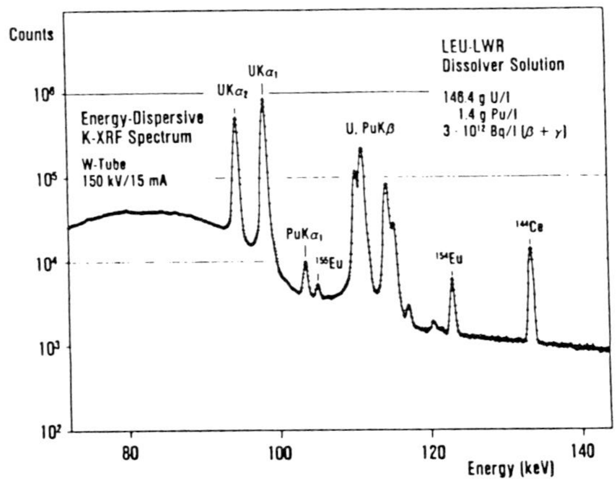

Fig. 7: Tube-excited XRF spectra from a LWR dissolver solution. Energy-dispersive K-XRF spectrum

A schematic plan of the present instrument installation at the European Institute for Transuranium Elements is shown in fig.8. The instrument is located outside a hot cell facility. It is connected to the interior of the hot cell through a tight stainless-steel tube, which serves for sample transport.

The two X-ray methods incorporated into the instrument use a single X-ray tube as photon source. The tube, operated at a potential of 150 kV and at tube currents of between about 10 and 15 mA, irradiates with two separate X-ray beams a dual sample consisting of a 2 cm glass cell for K-edge densitometry, and a 1 cm diameter polyethylene capsule for XRF. The straight-through X-ray beam for K-edge densitometry is highly collimated through a 12 cm long, 0.08 cm diameter tungsten collimator. The XRF-detector, which is located at the largest possible backward angle relative to the primary beam direction, views the XRF cell through a 14 cm long, 0.6 cm diameter tungsten collimator.

Both detectors are standard planar HpGe detectors. They are operated at total counting rates of between about 20 000 and 50 000 cps. The detector signals are digitized in two 400 MHz ADC's, and accumulated and analysed simultaneously on a computer-based data acquisition and analysis system.

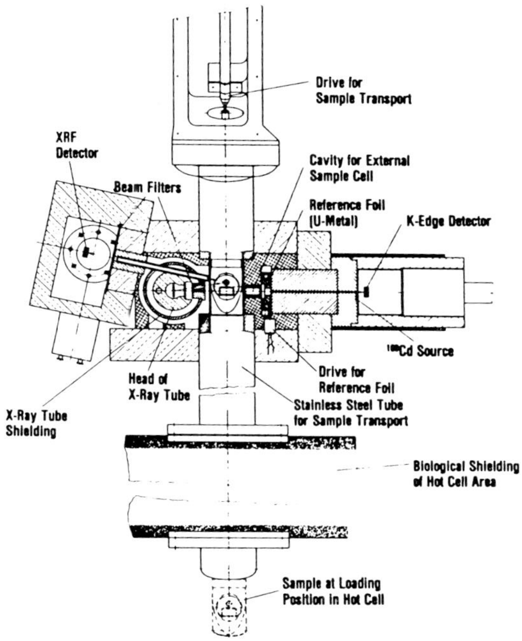

Fig. 8: Schematic plan of the hybrid K-edge/K-XRF instrument

5 CONCLUSIONS

A variety of new tools have been developed, which will allow for faster and more accurate analyses of the fissile nuclide masses, which enter and leave a reprocessing plant. Whereas the isotope correlation technique is an auxiliary technique, K-edge densitometry has been already introduced for routine output verification of uranium and plutonium. The automatic mass-spectrometric isotope dilution analysis method and K-edge densitometry combined with X-ray fluorescence will speed up the verification measurement considerably, when applied on-site.

LITERATURE

/1/ "The Isotope Correlation Experiment, ICE"
Edited by L. Koch, S. Schoof
ESARDA-2/81, EUR 7766 EN, KfK 3337(1981)

/2/ A. Giacometti
Proceedings of 6th Annual ESARDA Symposium, Verice ESARDA 17 p.213(1984)

/3/ "A Comprehensive Fuel Nuclide Analysis at the Reprocessing Input"
H.-J. Arenz, L. Koch, S. Schoof
Int. Symposium on Recent Adances in Nuclear Material Safeguards, IAEA, Vienna, 8-12 November 1982

/4/ "Einsatz eines vollautomatischen Massenspektrometers in der Spaltstoffflußkontrolle"
B. Brandalise, L. Koch, C. Rijkeboer, M. Romkowksi, W. Wilhelmi
EUR 5594 d, KfK 2319(1977)

/5/ "Automat zur Trennung von Uran und Plutonium aus Spaltprodukten durch Ionenaustauscher für die massenspektrometrische Analyse"
A. Bier, D. Bol, B. Brandalise, M. DeRossi, L. Koch
EUR 4917 d(1973)

/6/ "Automatic Separation of Uranium and Plutonium from Nuclear Fuels"
B. Brandalise, M. DeRossi, L. Koch, M. Wilker
Karlsruhe International Conference on Analytical Chemistry in Nuclear Technology, Karlsruhe, June 4-7 (1985)

/7/ P.DeBièvre et al.
Intern. Atomic Energy Agency, Vienna, Proceedings of Nuclear Safeguards Technology 1986, IAEA-SM-293/21 (1987)

/8/ "Reprocessing Output Verification by K-Edge Densitometry"
R. DeMeester, H. Eberle, S. Johnson, L. Koch, J. Michel-Piper, H. Nackaerts, H. Ottmar
Intern. Atomic Energy Agency, Vienna, Proceedings of Nuclear Safeguards Technology 1986, Vol.I, IAEA-SM-293/75, p.233-239 (1987)

/9/ "Field Demonstration of an X-Ray Densitometer for Uranium and Plutonium Input Verification in Reprocessing"
R. DeMeester, H. Eberle, L. Koch, E. Kuhn, H. Ottmar
Intern. Atomic Energy Agency, Vienna, Proceedings of Nuclear Safeguards Technology 1986, Vol.I, IAEA-SM-293/48, p.201-211 (1987)

Passivity of titanium and tantalum oxide films under irradiation

L. Elfenthal, J.W. Schultze

SUMMARY

Titanium, tantalum, zirconium, hafnium or niobium can be used as electrodes and materials for the Purex process. These valve metals build up a passive layer which gives good corrosion protection. The application may be limited by radiation induced corrosion effects, especially pitting and hydrogen embrittlement. The nuclear radiation effects of -particles and hot atoms (ha) can be simulated by ion implantation of He^+- and Pd^+-ions. Electrochemical and photoelectrochemical measurements have proved an amorphization and increasing electronic conductivity of the TiO_2-film after Pd^+-implantation. He^+-implantation causes low effects. All measurements give no hint of channel-like destruction or pitting.

Localized laser illumination was used to simulate electronic effects of radiation. The increasing laser power causes irradiation effects on TiO_2-films, especially crystallization and an oxide growth caused by light induced changes of the electric field.

L. Elfenthal*, Prof. Dr. J. W. Schultze, Institut für Physikalische Chemie II der Universität Düsseldorf, Universitätsstr. 1, D-4000 Düsseldorf, Federal Republic of Germany

* Part of the dissertation of L. Elfenthal

INTRODUCTION

The valve metals Ti, Ta, Zr, Hf and Nb are discussed as materials for electrodes and equipment in the Purex process for the recovery of plutonium and uranium from nuclear fuels. The metals ara passivated by an oxide layer of a few nm. This resistant passive film causes a good protection against corrosion. In the Purex process the materials are exposed to the influence of chemicals (products of radiolysis, aggressive ions, HNO_3) and the radiochemical influence (α, β, γ, n-radiation, hot atoms) which are shown schematically in Fig.1. The primary energy E_{prim} of radiation given is reduced within the saturation thickness d_s. The Purex solution causes moderation and the energy spectra of radiation hitting the surface range from E_{prim} to 0.

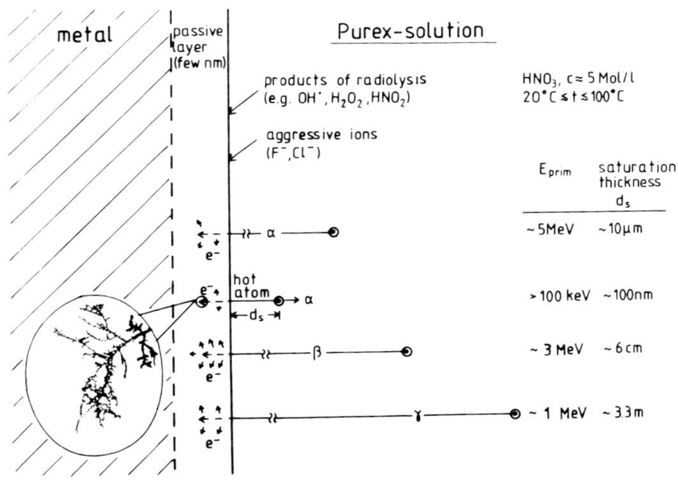

Fig. 1: Schematic diagram of the chemical and radiochemical influence of radiation on the passivated metal in contact with the Purex solution

Electrochemical processes at the electrodes and materials can be characterized by the current potential curves of anodic and cathodic processes in the Tafel-plot by i vs. ε as shown in Fig.2. The region of stability of passive layers and a possible shift of valve metals is limited by radiation induced corrosion or hydrogen embrittlement.

In order to simulate radiation effects the energy E and mass m of incident particles must be considered. Fig. 3 shows a double

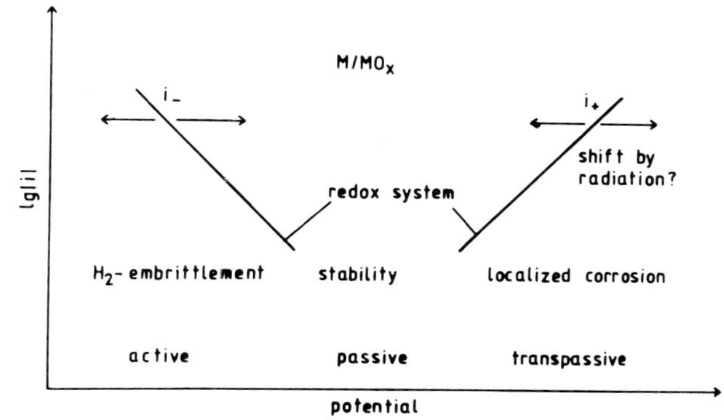

Fig. 2: Schematic diagram of the regions of electrode material stability and a lg i/ε plot (Tafel diagram) of a redox system with possible shifts of the curves under radiation as indicated.

logarithmic diagram of the relation of energy and mass for radiation and simulation experiments. The comparison shows the great variation of energy (20 - 200 keV) and mass (H.... Bi(Pu)) of ion implantation experiments. This means that α and hot atom simulation by ion implantation is possible.

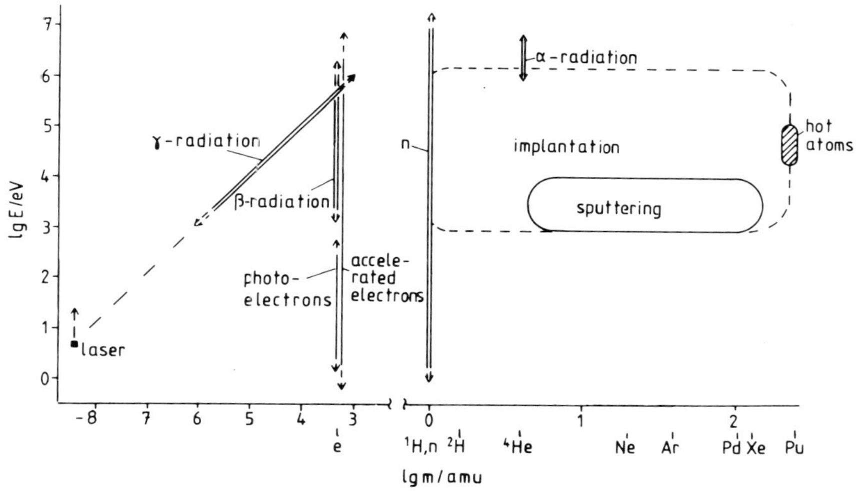

Fig. 3: Schematic mass-energy-diagram of radiation and simulation experiments

Our experiments of -simulation were carried out with ion implantation of He^+. Heavy ions (e.g. Pd^+ or Xe^+) were used to simulated hot atoms.

In the next stage the dose and overall energy of implantation must be considered. Table 1 shows data of radiation in the Purex process and in the simulation experiments. The materials for the equipment should be in contact with radioactive solutions for ten years or more: to get the same overall energy in 1 hr-simulation experiments we used the high dose rate of implantation which was five orders of magnitude higher than that of the Purex process.

TABLE 1: Data of radiation in the purex process and simulation experiments ($E_d = E_{prim} \cdot \dot{D} \cdot t/r_p$)

Irradiation	Primary energy E_{prim}	Max. range in passivated metals r_p	Dose rate $\dot{D}$/cm^{-2} s^{-1}	Absorption volume A/cm^{-3}	Irradiation time t/s	Integrated energy density in the absorption volume E_d/eV cm^{-3}
α	5 MeV	15 µm	10^7	$1.5 \cdot 10^{-3}$ *	$3 \cdot 10^8$	10^{25}
He^+-impl.	10-100 keV	80-490 nm	10^{12}	$8 \cdot 10^{-6}$-$4.9 \cdot 10^{-5}$ *	10^4	10^{25}
ha	100 keV	30 µm	10^6	$3 \cdot 10^{-6}$ *	$3 \cdot 10^8$	10^{25}
Pd^+-impl.	10-200 keV	8-65 nm	10^{12}	$8 \cdot 10^{-7}$-$6.5 \cdot 10^{-6}$ *	10^4	10^{26}
β	3 MeV	1-10 mm	10^{10}	0.1-1 *	$3 \cdot 10^8$	10^{26}
Laser-exp. (non-focussed)	4.8 eV, 0.1-10 mW	100 nm	10^{17}-10^{19}	$3 \cdot 10^{-8}$ **	1	$5 \cdot 10^{22}$-$5 \cdot 10^{24}$
Laser-exp. (focussed)	4.8 eV, 0.1-10 mW	100 nm	10^{21}-10^{23}	$3 \cdot 10^{-12}$ ***	1	$5 \cdot 10^{26}$-$5 \cdot 10^{28}$

* irradiated area 1 cm^2

** irradiated area $3 \cdot 10^{-3}$ cm^2

*** irradiated area $3 \cdot 10^{-7}$ cm^2

The following processes are important for the loss of energy of implanted ions:

- elastic collisions with target atoms: nuclear stopping (formation of Frenkel defects and defect cascades, especially for the interaction with heavy ions)

- non-elastic collisions with binding electrons of the target material: electronic stopping (electron excitation and ionization)

- elastic collisions with binding electrons of the target material (change of the ion charge)

- excitation of phonons.

The electronic effect of ions and radiation can be simulated by laser induced electron excitation. The laser photon energy is 4.8 eV (Fig.3). The variation of power (0.1 - 100 mW), irradiation time (t > 1 µs) and focussing (a few µm) of the laser yields a broad range of energy densities up to that of ß-radiation.

Fig.4 gives the time resolved energy flux during the interaction of high energy ions with target materials for radiation. Single energy transfer takes place during a primary time interval of 10^{-16} s. The projectile loses the energy in about 10^{-12} s. Fast energy dissipation takes place in the range of $10^{-12} < t < 1$ s. The possible effects in the time scale of measurements after implantation, $t > 10^2$ s, are diffusion of vacancies and interstitials, growth of passive film, corrosion and change of mechanical properties.

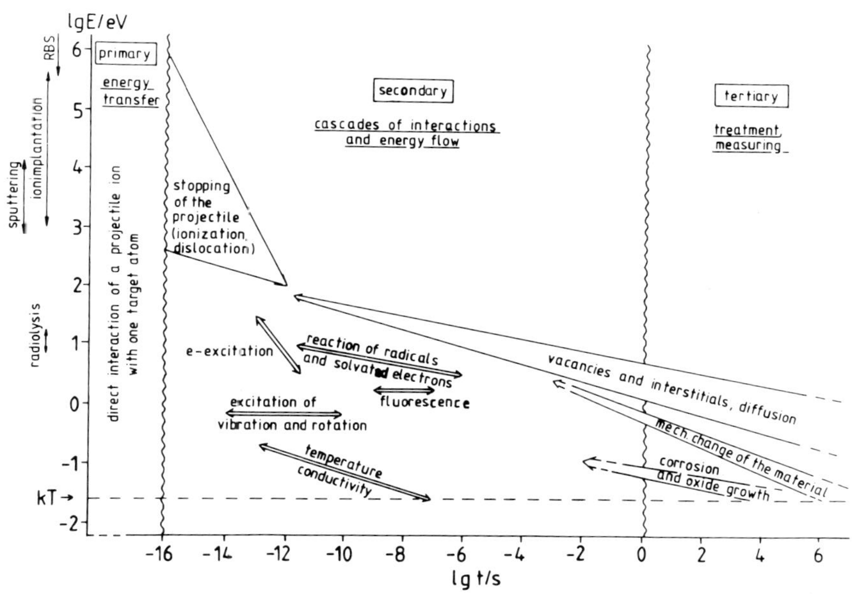

Fig. 4: Energy flow dependent on time due to the interaction of an accelerated ion with material

2 EXPERIMENTAL

The titanium electrodes (purity 99.6%) were electropolished in an aqueous acetic acid-perchloric acid mixture. This procedure yields a surface which is almost free of oxide (d_{ox} < 1 nm) with an assumed roughness factor of 1. Oxide films were formed galvanostatically in 1 N H_2SO_4 (14 mA cm^{-2}) by anodic polarization up to 20 V, corresponding to a film thickness of 40 nm /1/. The thinner oxide films below a formation potential of 20 V are amorphous. Above this potential X-ray refraction shows anatase modification /2/. Implantation of the electrodes with 200 keV Pd^+-ions or 100 keV H^+-ions yields a concentration profile which is shown in Fig.5. The dose of implanted He^+-ions is $1.6 \cdot 10^{16}$ cm^{-2} (~4 at -%) and $8 \cdot 10^{16}$ cm^{-2} (~17 at -%) and that of implanted Pd+-ions is $2.6 \cdot 10^{15}$ cm (~1 at -%). The results of the measurements were compared with the experimental data of an unimplanted electrode (NOT).

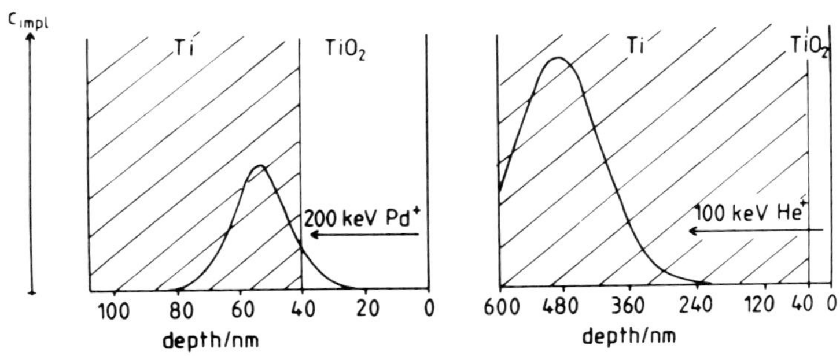

Fig. 5: Schematic depth profile of the implanted ion (different scales)

Electrochemical experiments were carried out in pure deaerated 1 N H_2SO_4. For the ETR-measurements the $Fe^{2+/3+}$ redox couple was added with a concentration of 0.05 M. All potentials are referred to the standard hydrogen electrode (SHE). For a detailed description of the experimental techniques see /3-5/.

For laser experiments TiO_2 films were prepared by polarization between 7 V and 100 V, coresponding to a film thickness of 14 to 200 nm. The tantalum electrodes were mechanically polished and the 20 nm Ta_2O_5-layer was formed by polarization up to 10 V.

3 SIMULATION OF RADIATION BY α-PARTICLES AND HOT ATOMS

The electronic and electrochemical behaviour of passive films changes by implantation. The current during anodic polarization after implantation can be correlated to the migration of defects and repassivation /3/. Capacity measurements yield information about the charge distribution in the oxide film, the donor density N and the dielectric constant D /6, 7/. Electron transfer measurements prove the electronic conductivity of the oxide film and are very sensitive to electronic states around the Fermi level E_F /4,7/. Photoelectrochemical experiments yield information about the band structure and the crystallinity of the oxide /4, 6, 7/.

3.1 ANNIHILATION OF RADIATION EFFECTS BY ANODIC REPASSIVATION

Radiation with ions of oxide films produce Frenkel defects, e.g. titanium interstitial ions and vacancies:

$$Ti_{Ti} \rightarrow Ti_i^{4+} + V_{Ti}^{''''}$$

The effectivity of this process depends on the energy and mass of the implanted ions. The following anodic polarization causes a defect migration at a field strength much less than that of the oxide formation /3/. Fig. 6 shows the first potentiodynamic cycle of implanted and not implanted electrodes. In the potential range from $\varepsilon = 0$ to 2 V partial current densities of capacitive charging i_c, oxygen evolution i_o, corrosion and other side reactions i_x and oxide formation i_{ox} may contribute to the total current density i:

$$i = i_c + i_o + i_x + i_{ox} \qquad (2)$$

i_c, i_o and i_x are always smaller than 1 uA cm^{-2}. This is shown by the very low current density of the unimplanted oxide film NOT and the decrease of the current density of the Pd^+-implanted film in the following repassivation cycle (Fig. 6 shows the first and second potentiodynamic cycle).

The irreversible anodic current density of the 1 atom-% Pd^+-implanted electrode during the first cycle corresponds to a defect annihilation

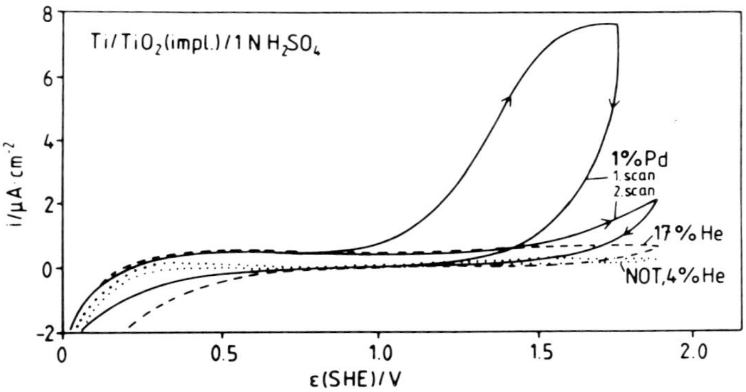

Fig. 6: Repassivation during the first potentiodynamic cycle ($d\varepsilon/dt$ = 10 mV/s) for different implanted and not implanted electrodes

(re-organisation) and new oxide formation at the oxide/electrolyte interface by Ti_i^{4+} (Fig. 7):

$$Ti_i^{4+} + 2H_2O \longrightarrow TiO_2 + 4H^+ \qquad (3)$$

and filling of vacancies at the metal/oxide interface /3/:

$$V_{Ti} + Ti_m \longrightarrow Ti_{Ti} + 4e^- \qquad (4)$$

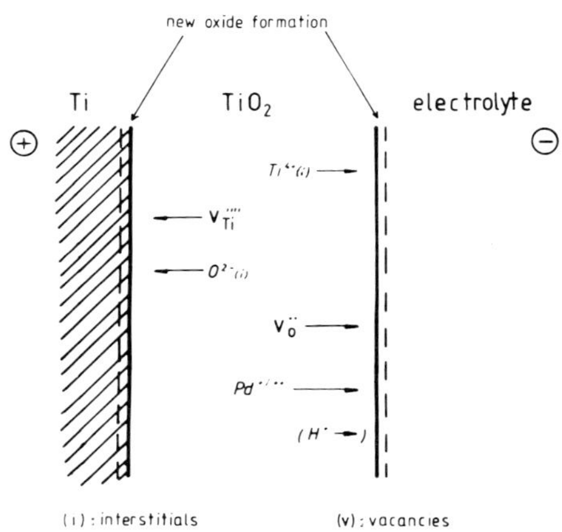

Fig. 7: Migration of different defects to the oxide interfaces during anodic polarisation

Implantation of 17 atom-% He^+ yields a very small increase of current density and implantation of 4 atom-% He^+ shows the same negligible current as the blank electrode NOT.

The measuring of repassivation current is sensitive for defects caused by ion implantation /3/. Therefore, Fig. 6 demonstrates successfully that 1 atom-% Pd^+-ions produce large amorphization but the effect of He^+-implantation (4 and 17 atom-%) is negligible. The experiments have not given any indication of channel-like destruction or pitting but a continuous repassivation of the oxide layer.

3.2 OXIDE PROPERTIES FROM ELECTRODE CAPACITY

Further information on possible radiation effects can be obtained from the electrode capacity which is a measure of the defect concentration and dielectric properties /7/. TiO_2-films show a potential dependence of the capacity like an n-type semiconductor. In the region of low potentials above the flat band potential ($\varepsilon_{Fb} \approx 0$ V) the Mott-Schottky-equation /8,9/ may be valid:

$$1/C^2 \approx 1/C_{sc}^{\;2} = 2/eDD_oN\ (\varepsilon - \varepsilon_{Fb} - KT/e) \qquad (5)$$

D and D_o are the dielectric constant of the oxide and the vacuum and N stands for donor concentration within the passive film. The slope of the $1/C^2$ -ε-curve at low potentials gives the product N·D. At higher potentials the capacity becomes constant if the oxide thickness d_{ox} is the same as the thickness of the space charge layer d_{sc} . In this case the condensor formula should be valid /9/:

$$C = D \cdot D_o/d_{ox}$$

It is well known that amorphization by ion implantation yields an increase of C caused by an increase of N and D /3-6/.

Fig.8 shows the capacities of a blank (NOT) and various implanted films in dependence on ε during the cathodic sweep. The unimplanted electrode has a capacity of about 1 $\mu F\ cm^{-2}$. Implantation of Pd+ increases the capacity by 1 order of magnitude ($C \approx 12\ \mu F\ cm^{-2}$) but implantation of He+ yields only a small increase in the potential range between 0 and 0.2 V (shift of the flat band potential). Table 2 gives the values of N, D, E_g and the implantation concentration N_{impl}.

Electrode	$N_{impl}/10^{-19}cm^{-3}$	$N/10^{-19}cm^{-3}$	D	E_g/eV
1% Pd	88	4.6	452	3.4
4% He	352	0.12	45	3.3
17% He	1496	0.1	45	3.27
NOT	-	0.1-0.2	32-45	3.15

Table 2: Implantation concentration N_{impl}, donor concentration N, dielectric constant D and band gap energy E_g (from photocurrent measurements) for various implanted and blank electrodes

The ratio $N \ll N_{impl}$ shows that the Pd-ions do not act as donors and we conclude that all occupied electronic levels of the implanted ions are lying deep in the band gap or they are localized by opposite charges. The increase of N and D after Pd^+-implantation indicates a distinct amorphization of the passive film. The effects of He^+-implantation are negligible.

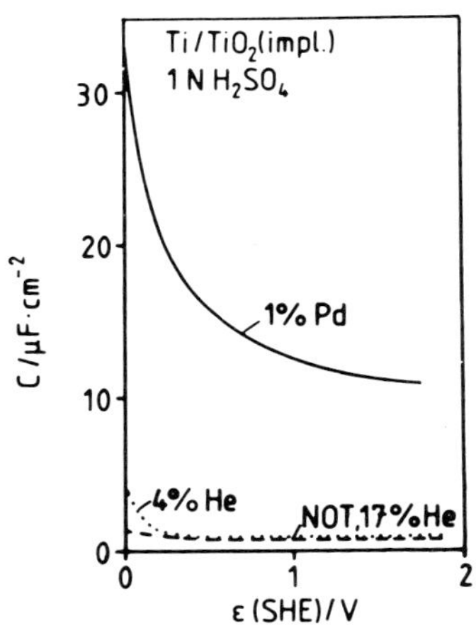

Fig. 8: Electrode capacity C during the first potentiodynamic cycle (cathodic sweep, $d\varepsilon/dt$ = 10 mV/s) for different implanted and not implanted electrodes (cathodic sweep)

Repassivation causes a decrease of N and D /3/. Hence, these defects can be annihilated also.

3.3 ELECTRONIC PROPERTIES OF OXIDE FROM PHOTOELECTROCHEMISTRY

The energetic distribution of the radiated oxide films can be derived from photocurrent measurements /10/. The photocurrent spectra of the implanted and unimplanted TiO_2-films show an absorption threshold at photon energies $h \cdot \nu$ of about 3.2 eV (Fig. 9). Implantation can cause an increase as well as a decrease of the photocurrent i_{ph} depending on the oxide, the implanted species and concentration. In this case the absolute i_{ph} will not be discussed, but the important result is the band gap energy E_g or in the case of an amorphous semiconductor the mobility gap energy E_m /11/. The results of the photo-measurements are given in Table 2.

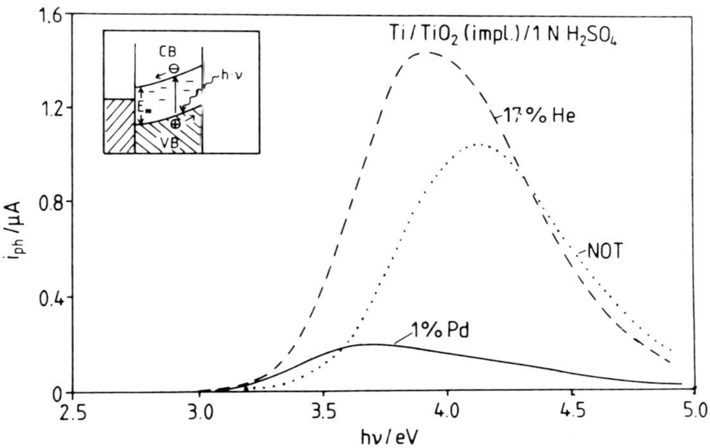

Fig. 9: Photocurrent spectra at ε = 1.5 V for various implanted and not implanted electrodes. The principal mechanism of photoinduced electron excitation is shown in the insert

Increasing amorphization of the oxide by ion implantation causes localized sub-band states near the band edges and the exponential band tails into the band gap (Fig. 10). Therefore, the mobility gap energy decreases and the density of states between valence and conduction band increases /10/.

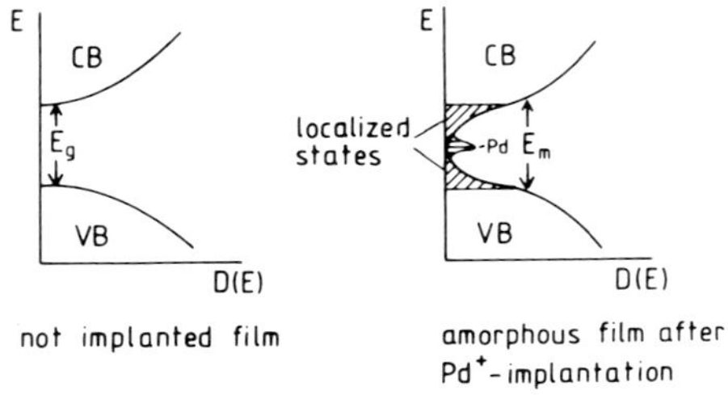

Fig.10: Schematic diagram of the density of states for an Pd^+-implanted TiO_2-film

3.4 APPLICABILITY OF VALVE METALS AS ELECTRODE MATERIALS FOR ETR

At unimplanted TiO_2 films cathodic electron transfer reactions (ETR) take place at high overvoltages, but anodic reactions are blocked, i.e. currents are < $1 \mu A/cm^{-2}$ (Fig.11, NOT). This can be explained by the high band bending at anodic potentials /12, 13/. Cathodic and anodic ETR are enhanced by several orders of magnitude after Pd+-implantation (Fig.11). The increase of currents after He+-implanation is even smaller. This can be explained by Pd and defect states lying close the Ferme level. Resonance tunnel processes via these localized defect states can easily occur if their distance d_r < 1 nm. Other states $E_{def} \gg E_F$ due to radiation induced defects are less effective for ETR /14/; they are observed in the capacity measurements and located near the conduction band.

This experiment shows the improved conductivity after ion implantation. This is a most important fact for the corrosion behaviour of Ti in the Purex process, since radiation will increase the conductivity and decrease the overpotential η of electrode reactions.

Experiments by Kaesche and Kaiser show the behaviour of Ti in HNO_3 /15/. At cathodic potentials a dangerous hydrogen embrittlement was observed. Measurements in the simulated Purex solution show a depositon of catalytically active noble metals (Pt, Pd, Ru) on Ti-cathodes. The dendrite-like structure of these metals is formed slowly, which diminishes the overpotential of the reduction processes by some hundred mV (Fig. 12) /16/. Radiation induced amorphization and increasing conductivity will cause an even faster adjustment of the

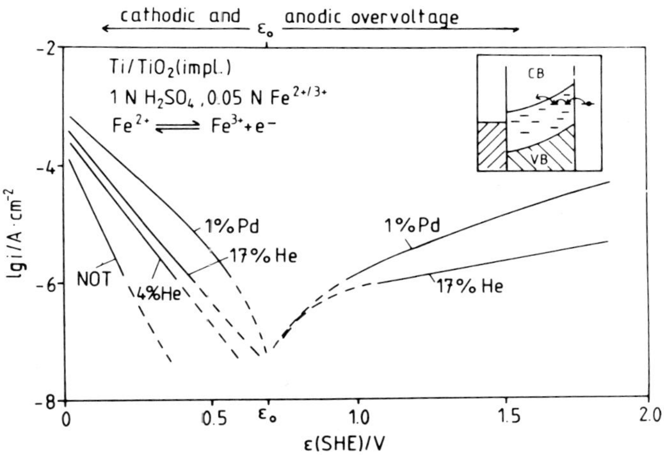

Fig.11: Tafel plots of the $Fe^{2+/3+}$ reaction on Pd^{+}- and He^{+}-implanted and not implanted electrodes ($d\varepsilon/dt$ = 10 mV/s, cathodic sweep). The principal mechanism of ETR is shown schematically in the insert

equilibrium potential of the electrolyte. Hence the potential ε will shift even more quickly the positive region where Ti is stabilised by the passive film.

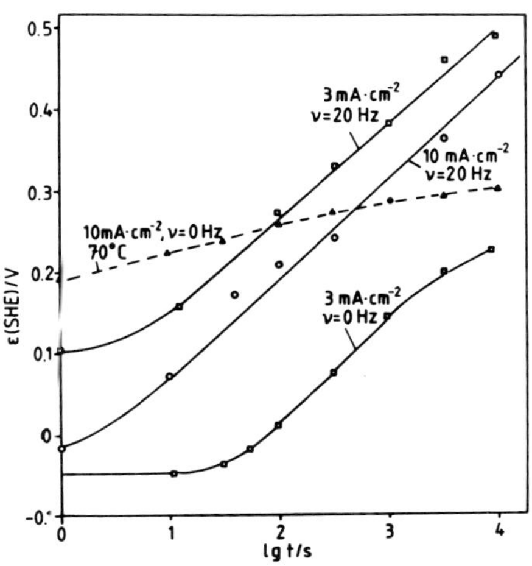

Fig.12: Shift of Ti-potential with time under technical conditions measured with a disc electrode (i = const., 3 M HNO_3, 1 M $UO_2(NO_3)_2$, 0.005 M Ce^{3+}, 0.003 M Pd^{2+}, 0.001 M Ru^{3+}). 25° C (full line) and 70° C (dashed line) /16/

3.5 COMPUTER CALCULATION OF RADIATION EFFECTS

The interaction of high energy ions with solids can be calculated by a computer programme /17/. Fig. 13 shows the trace of the implanted He^+ and Pd^+-ions. The heavy Pd^+-ion yields many defects and recoil cascades in contrast to the He^+-ion. The recoil distribution is shown in Fig. 14. The number of displaced O-ions is twice as high as that of the heavier Ti-ions. Table 3 gives a quantitative interpretation of these results and shows the dissipation of the implantation energy by different mechanisms of energy loss: ionization, Frenkel-defects and phonons. In the case of He^+-implantation 2 % of the total implantation energy generates Frenkel-defects but implantation of Pd^+ leads to a number of defects which is higher by 1 order of magnitude. This is the reason for the higher amorphization after Pd^+-implantation. These calculations are consistent with our experiments and support the physical picture derived before.

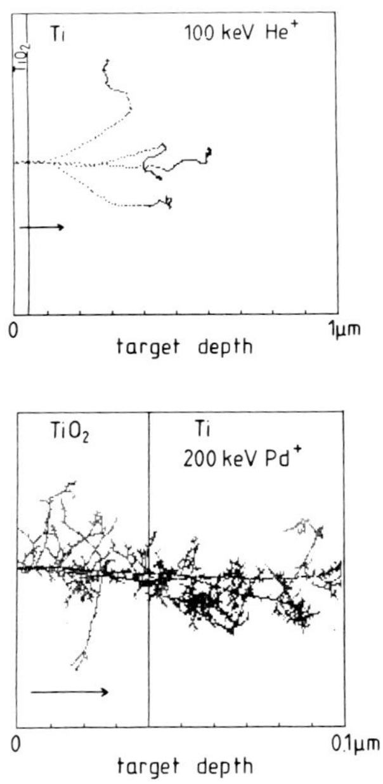

Fig.13: Computer simulation of the interaction of Pd^+- and He^+-ions with a 40 nm TiO_2-film on Ti (different scales) /17/

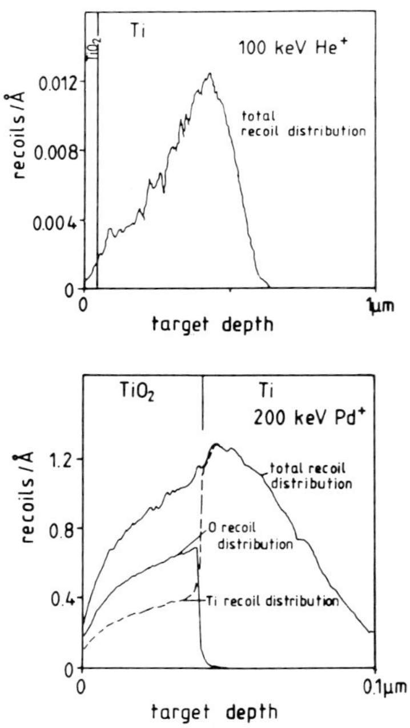

Fig.14: Computer simulation of the recoil distribution after Pd^+-and He^+-implantation (40 nm TiO_2 on Ti, different scales) /17/

TABLE 3: Dissipation of the implantation energy by different energy loss mechanisms for implantation of Pd^+- and He^+-ions into a 40 nm TiO_2 layer on Ti

Ion concentration /atom-%	Total implantation energy /eV cm^{-2}	Loss of energy/%			Total integrated energy loss for Frenkel defects/eV $Å^{-1}$	Total defect number for ion and recoils /vac ion^{-1}
		ionization	Frenkel-defects	phonons		
4 % He^+	$1.6 \cdot 10^{21}$	94	2	4	0.44	50
17 % He^+	$8 \cdot 10^{21}$					
1 % Pd^+	$5.2 \cdot 10^{20}$	37	26	37	84	6240

4 SIMULATION OF ELECTRON EXCITATION BY LASER EXPERIMENTS

Table 3 has shown that ionization is an important mechanism for loss of energy for α-radiation and less for hot atom radiation. Ionization and electron excitation are most important effects for ß- and γ-radiation. The corresponding generation of pairs of electron holes can be simulated by localized laser experiments. The principle of the electron excitation was shown in the insert of Fig.9.

An Ar^+-laser with a wavelength of 514 nm $\hat{=}$ 2.4 eV was used. The light passed a frequency doubler and the photon energy was raised to 4.8 eV. The power output was in the range between 0.1 and 100 mW. The electrochemical cell was mounted on a computer controlled x-y stage which could be moved in steps of 1 µm. For a detailed description of the experimental techniques see /18/.

Thin TiO_2-films d < 20 nm are yellow in colour and homogeneous. The colour of thicker oxide films of about 40 nm shifts to blue and the films show a distinct grain structure with various colours due to the polycrystalline underlying metal. The size of the grains is about 10 um. The localized photocurrent measurements by a scanning laser beam with different thicknesses of oxide can reveal the photo-response of the different grains (Fig.15). The photocurrent of the thin TiO_2-film (ε_{ox} = 27 V, $d_{ox} \approx$ 54 nm) can be correlated to the grain structure. Implantation of Pd+ equalizes the passive layer (ε_{ox} = 20 V, $d_{ox} \approx$ 40 nm) and the photo-response is fairly homogeneous.

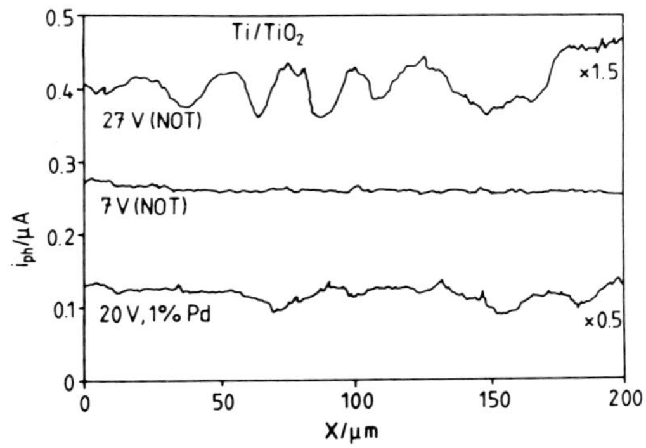

Fig.15: Analysis of not implanted (7 V≙ 14 nm, 27 V≙ 54 nm) and Pd^+-implanted (20 V≙ 40 nm, 1% Pd) TiO_2-films by microprobe photocurrent measurements (ε = 1.5 V, light power P ≈ 0.1 mW, 1 N H_2SO_4)

The different photocurrents of various grains can be explained by the different thicknesses of oxide or by changes in the electronic structure. Sub-band states can act as traps and can, therefore, influence the recombination rate of the photocurrent.

Radiation effects by laser illumination depend on the light density P, the illumination time t, the oxide thickness d_{ox} and the electrode potential ε. Table 1 shows the wide range of dose rate for laser experiments. For detailed investigations homogeneous 7 V TiO_2 ($d_{ox}\approx$ 14 nm) and 10 V Ta_2O_5 ($d_{ox} \approx 20$ nm) films were illuminated. The influence of ε was investigated in the "locodynamic" experiment: the electrodes were scanned by a laser beam and the electrode potential was increased simultaneously during the scan. Hence, the radiation effect is seen on the electrode in dependence on ε. Fig.16 shows the scan line on the TiO_2-film. Radiation induced effects, colour changes and crystallization of the films starts at potentials above 0.5 V and becomes more intensive and laterally extended at higher potentials. Similar results were obtained in another experiment. A TiO_2-film was successively scanned by a laser beam at constant ε. The photocurrent decreased with the increasing number of scans and showed a continuous increase of inhomogeneity (Fig.17).

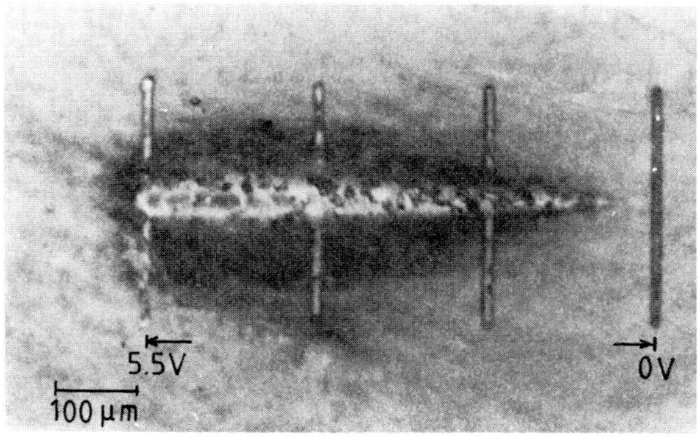

Fig.16: Photo of a laser scan line during increasing anodic potential on a TiO_2 film (d = 14 nm, $\mathcal{E}$ = 0-5.5 V, P ≈ 1 mW, d /dt = 18 mV/s, 1 N H_2SO_4)

These radiation effects are induced by a change in the local field distribution by laser illumination. Defects become mobile especially in the region of the space charge layer and transfer reactions in the double layer are probable. This yields an increasing oxide thickness

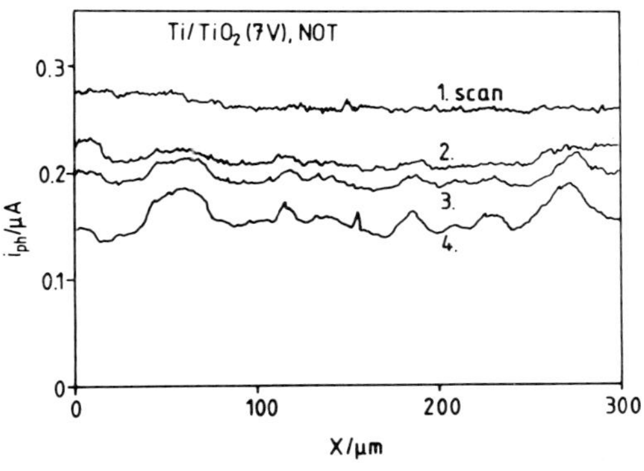

Fig.17: Successive scans on a TiO_2 film (d = 14 nm,$\mathcal{E}$= 1.5 V, P ≈ 0.1 mW, 1 N H_2SO_4)

and a variation in the electronic structure such as sub-band states due to defects.

Light induced changes of the oxide film are not due to an effect of temperature. Calculations of the local heating ΔT < 10° have shown that T remains far below the temperature of crystallization of amorphous TiO_2, T_c = 480° C /19/, which is not reached by the laser illumination. A determining temperature effect could not explain that the crystallization and oxide growth are potential dependent and vanish at $\mathcal{E}$ < 0,5 V. The "locodynamic" experiment at a Ta_2O_5-electrode shows no crystallization. Around the laser trace regions of rainbow colours are seen at potentials higher than 0.5 V. This indicates different thicknesses of the oxide layer (interference colours) /18/.

Inhomogeneous layers of TiO_2 were measured by Bullock and Butler /20, 21/. Our experiments correlate localized photoelectrical investigations of thin passive films with the grain structure, oxide growth and laser induced radiation effects.

5 CAN RADIATION INDUCE LOCALIZED CORROSION ?

Thus far the implantation experiments demonstrated the amophization and the absence of local corrosion. An amorphous film seems to cover the metal even better than a crystalline one with larger grain boundaries. In order to make certain of this point we investigated the passivation of a fresh surface by simulating a hypothetic, unobserved pit induced by radiation. This was followed by pulse measurments. Fig.18 shows the current density of an almost bare Zr-electrode after a potential step from 0 V to $\mathcal{E}_p$ = 2 V. After a first peak the current density decreases continuously while localized corrosion would increase the current. For 1 N HNO_3 the current was even 2 orders of magnitude lower than for 1 N H_2SO_4 in the first 10^{-3} s of the experiment. The oxidation of HNO_3 built up a ZrO_2-layer of a few nm before the experiment started.

This means that even in the case of localized corrosion a pit would be repassivated quickly. Thus, in all these experiments we did not find any evidence of instability induced by radiation.

6 CONCLUSION

The simulation of radiation effects by ion implantation is complementary to "hot" experiments /22/. With the implantation technique it is possible to separate single effects of irradiation. The experiments show the changes in electrochemical and photoelectrochemical behaviour of passive films, especially the corrosion behaviour.

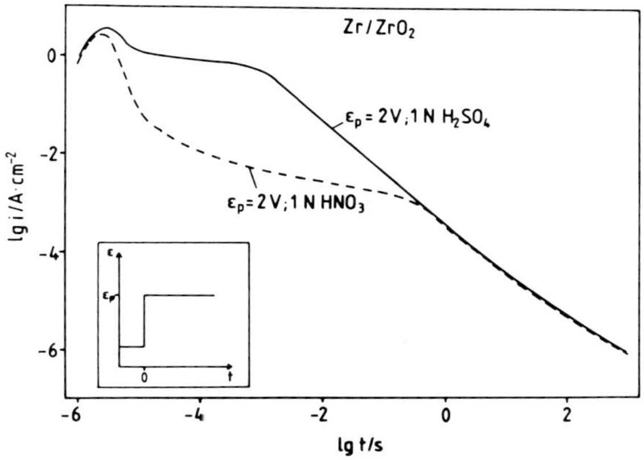

Fig.18: Potentiostatic pulse measurements on Zr from $\varepsilon = 0$ V to $\varepsilon_P = 2$ V in 1 N H_2SO_4 and 1 N HNO_3

The radiation damage caused by Pd^+-implantation in TiO_2-films has been demonstrated successfully by measurements of repassivation, capacity, ETR and photocurrents. Amorphization decreases with anodic polarisation, as reflected in repassivation, in the dielectric constant D and also in the photoelectrochemical measurements. Mobile defects (Frenkel-defects) migrate to the surface and build up new oxide layers.

Amorphization yields an increasing electronic conductivity and causes a fast adjustment of the equilibrium potential of the electrolyte (stabilization of the oxide layer). He^+-implantation causes a very low amorphization. Theoretical computer calculation shows that a great part of the implantation energy of He^+-ions takes place in ionization processes. Therefore, the radiation effect of hot atoms exceeds that of α-particles by far.

The surface structure of thin oxide layers on Ti has been analysed by microprobe photocurrent measurements. Changes of the photoresponse form one grain to another could be explained by a different thickness of the passive layer and a varying electronic structure.

Laser radiation of TiO_2-films changes their colour and crystallinity. The photo-generated pairs of electron holes vary the electric field within the oxide and the increasing migration yields a varying defect structure and an increasing thickness which cannot be explained by the effects of temperature.

ACKNOWLEDGEMENT

The support for this work by the Bundesministerium für Forschung und Technologie is gratefully acknowledged.

LITERATURE

1. K. Leitner, J.W. Schultze, U. Stimming, J. Electrochem. Soc. 133 (1986) 1561

2. L. Arsov, M. Froelicher, M. Froment, A. Hugot-Le Goff, J. de chimie physique 72 (1975) 275

3. L. Elfenthal, K. Leitner, J.W. Schultze, O. Meyer, Ber. Bunsenges. Phys. Chem. 91 (1987) 432

4. J.W. Schultze, L. Elfenthal, K. Leitner, O. Meyer, J. Mat. Sci. and Eng., in press

5. Leitner, L. Elfenthal, J.W. Schultze, O. Meyer, International Symposium on Trends and new Applications in Thin Films, Strasbourg March 1987, Supplement of "Le Vide les couches minces" 235 (1987) 491

6. B. Danzfuß, J.W. Schultze, U. Stimming, O. Meyer, in "Passivity of Metals and Semiconductors", ed. by M. Froment, Elsevier Science Publishers B.V., Amsterdam, p. 503

7. J.W. Schultze, B. Danzfuß, O. Meier, U. Stimming, Mat. Sci. and Eng. 69 (1985) 273

8. K. Bohnenkamp, H.J. Engell, Z. Elektrochem. 61 (1957) 1184

9. J.W. Schultze, U. Stimming, J. Weise, Ber. Bunsenges. Phys. Chem. 86 (1982) 276

10. Danzfuß, J.W. Schultze, U. Stimming, in "Grundlagen von Elektrodenreaktionen", (Ed. J.W. Schultze), Dechema-Monographien Bd. 102 (1986) 465

11. R.H. Bube, "Photoconductivity of Solids", Wiley, New York, 1967

12. J.W. Schultze, L. Elfenthal, J. Electroanal. Chem. 204 (1986) 153

13. W. Schmickler, J.W. Schultze, Mod. Asp. of Electrochem. 17 (1986) 357

14. W. Schmickler, U. Stimming, Thin Solid Films 75 (1981) 331

15. H. Kaesche, H. Kaiser, private communication

16. J.W. Schultze, A. Borgerding, Dechema-Monographie, in press

17. Computer program "TRIM 86", J.P. Biersack, Berlin

18. K. Leitner, J.W. Schultze, in preparation

19. H.M. Naguib, R. Kelly, Rad. Eff. 25 (1975) 1

20. K.R. Bullock, G.M. Trishan, R.G. Burrow, J. Electrochem. Soc. 130 (1983) 1283

21. M.A. Butler, J. Electrochem. Soc. 130 (1983) 2358

22. G. Marx, in "Grundlagen von Elektrodenreaktionen" (Ed. J.W. Schultze), Dechema-Monographien Bd. 102 (1986) 255

Corrosion of valve metals under the influence of various components including plutonium

G. Marx, A. Bestanpouri, W. Schönemann, D. Wegen

SUMMARY

The corrosion behaviour of the valve metals Ti, Zr, Hf, Nb and Ta as investigated under the practical conditions of the Purex process by use of the radioisotope method (RIM), a combination of classical electrochemical methods with neutron activation analysis, allowing to determine even very small quantities of corroded material.

Special attention was paid to the influence of Pu on the corrosion of the materials aforementioned but also the effects of U and halides were taken into consideration.

Potentiostatic and potentiodynamic measurements were performed not only in the anodic but also in the cathodic range including those at rest potentials, the HNO_3 concentration ranging from 0.5 to 8.0 mol/l and the temperature from 20°C to 60°C.

The results obtained for the annual corrosion rates show Ta, Nb and Hf to be resistent towards corrosion under the practical conditions of the Purex process whereas Ti and especially Zr have minor qualities.

Prof. Dr.-Ing. G. Marx, Dr. A. Bestanpouri, W. Schönemann, D. Wegen; Freie Universität Berlin, Institut für Anorganische und Analytische Chemie, Forschungsgruppe Radiochemie, Fabeckstr. 34-36, D-1000 Berlin 33, Federal Republic of Germany

1 INTRODUCTION

Valve metals are those elements, which form an oxide layer, when anodically polarized, as for instance Ti, Zr, Hf, Nb, Ta. Since this oxide layer resists corrosion, valve metals are widely used in industries especially in the field of spent fuel processing. So dissolvers and evaporators are made of Zr and Ti and those apparati developed for applying electrochemical procedures to the Purex process in order to minimize the waste output [1]. Although the corrosion behaviour of valve metals was already investigated in detail in the past there still has been a lack of knowledge with respect to their applicability to practical conditions as for instance in aqueous nitric acid solutions and especially in the presence of plutonium. The destruction of protecting oxide layers by α- radiation was widely suspected especially in the case of HfO_2 which crystallizes nearly in the same lattice as PuO_2. Hafnium has been suggested to be preferred for equipping mixer settlers and pulsed columns with electrochemical units for oxidation and reduction in order to prevent criticality [2].

Therefore the corrosion of the 4th and 5th group elements Ti, Zr, Hf, Nb and Ta had to be investigated systematically under practical conditions of the Purex process, which means in various nitric acid solutions (0.05 - 5 mol/l) acting as sophisticated redox systems, moreover at various temperatures (20°C, 40°C, 60°C), in the presence of halides (F-, Cl-) and uranium and last but not least in the presence of plutonium at concentration of 2.2 g/l and more in order to contribute to the optimization and safety of spent fuel processing.

2 EXPERIMENTAL DETAILS

All experiments were performed by use of the radioisotopic method (RIM), described in detail elsewhere [3-7]. RIM combines electrochemical procedures as potentiostatic and potentiodynamic measurements, including also the determination of impedance, with neutron activation and γ-spectroscopy performed with Ge(Li) detectors and multichannel analyzers (MCA).

Neutron activation analysis makes even traces of metals dissolved by corrosion to be determined γ-spectroscopically, allowing the partial current density for corrosion to be separately calculated and compared with the overall current density, determined electrochemically, the difference being the current density which forms the oxide layer.

Since the thickness of this layer can also be calculated from impedance measurements, even side reactions, which may occur at the electrode, especially in aggressive redox systems like HNO3 can be determined from any deviation of the overall current density from the sum of the current densities, which result from the formation of the oxide layer and the dissolving of the material under investigation.

In table 1 the nuclear reactions for activating the valve metals are listed up. It can be seen that even nonisotopic tracers produced by (n;p) reactions are used for labelling, the applicability of this procedure being proved by various experiments (fig.1).

Hf-174(n,γ) Hf-175 $\rightarrow$ Ec $\rightarrow$ Lu-175
Hf-180(n,γ) Hf-181 $\rightarrow$ β^- $\rightarrow$ Ta-181

Ti-46(n,p) Sc-46 $\rightarrow$ β^- $\rightarrow$ Ti-46
Ti-47(n,p) Sc-47 $\rightarrow$ β^- $\rightarrow$ Ti-47

Zr-94(n,γ) Zr-95 $\rightarrow$ β^- $\rightarrow$ Nb-95 $\rightarrow$ β^- $\rightarrow$ Mo-95

Ta-181(n,γ) Ta-182 $\rightarrow$ β^- $\rightarrow$ W-182

Table 1: Nuclear reactions used for neutron activation

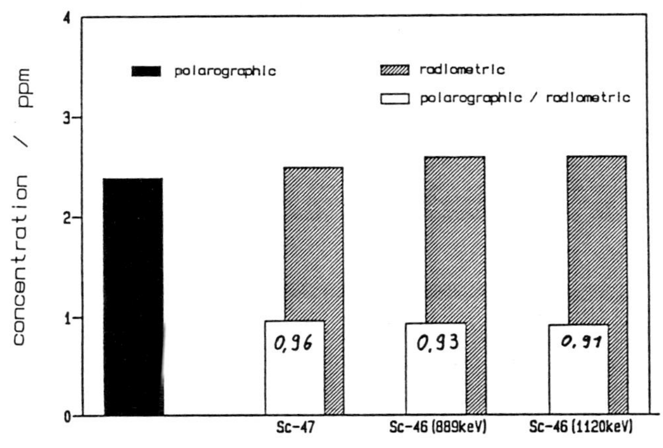

Fig. 1: Comparison of polarographic with radiometric methods

The results for the analysis of Ti polarographically obtained, for instance, confirm those got from radiochemical methods within the error limits, which must be taken into consideration especially with respect to the low concentrations used.

In addition the analysis of stainless steel performed by use of a variety of analytical methods also underlines the reliability of RIM, even in case of applying nonisotopic tracers.

3 CORROSION OF HAFNIUM UNDER PUREX CONDITIONS

3.1 EXPERIMENTS ON HAFNIUM IN THE ANODIC RANGE

Hafnium forms anodically a non-conducting oxide layer which blocks all charge tranfer reactions. It is an insulator up to a voltage of more than +80 V. Local corrosion which may happen at approximately +3 V can be prevented by longterm polarization for at least 20 h or even 100 h just below the potential of local corrosion. In this case even a potential of +80 V does not increase any current density or corrosion rate.

The annual corrosion rates of some μm/a are very small as can be seen from table 2. They are even decreased by adding plutonium, which homogeneously forms mixed crystals of $(Hf;Pu)O_2$, this result being proved by XPS measurements [3,8].

Hf Corrosion Rates in μm/a

Electrolyte Solution	c_{HNO_3}	techn. Hf (97% KfK) Maximum Value	techn. Hf (97% KfK) Mean Value	Pure Hf (99.5%, HEK) Maximum Value	Pure Hf (99.5%, HEK) Mean Value
Pure Nitric Acid	0.5 M	71	51	67	67
	1-5 M	17	8	7	4
With Addition of 2.2 g/l Pu(IV)	0.5 M	7	7	9	7
	1-5 M	14	6	5	3
With Addition of 2.2 g/l Pu(III) 0.2 g/l Hydrazine	0.2-1 M	2	2	2	2

Long-term Potentiostatic Polarization at +2 V
T = 20 °C

Table 2: Compilation of corrosion results for hafnium

Pu^{3+} and PuO_2 on the other hand are not incoporated into the oxide layer. With respect to the Pu content of the oxide layer three radiation effects must be taken into account [9]:

i: the radiolytic effect
ii: the photoelectric effect
iii: the destructive effect

The radiolytic effect is an event within the field of radiation chemistry. It produces H_2O_2, causing a potential shift towards higher values. This effect is important for thin oxide layers because in this special case the energy absorption within the oxide layer can be neglected.

The photoelectric effect influences the characteristics of the boundary semiconductor / electrolyte solution by forming additional charges, which significantly increase the minority group of the special semiconductor under consideration. This effect shifts the potential towards positive values for p-semiconductors and vice versa, under the condition that the oxide layer is sufficiently thick.

Both the effects just mentioned are only effective for the lenght of the time the material is exposed to radiation.

The destructive effect produces Fraenkel disorders within the oxide layer for a certain time, their decrease being caused by annealing.

These effects can be investigated by in situ measurements, especially by use of impedance and photoelectric methods.

In general they are observed at flux densities of approximately 1012 $cm^{-2} \cdot s^{-1}$. Since the Pu content of the oxide layer is only 1.5 atom%, which correponds to a desintegration rate of 100 Bq, no effects of this kind can be detected in this special case.

In literature simulating experiments have been described which were not performed under in situ conditions, being unreliable therefore [9].

Bode plots obtained from pure HfO_2 layers and from those containing plutonium have different slopes, showing the influence of Pu on the physical qualities of the oxide layer [3].

From measurements of the capacity no dependence of the thickness of the Ti and Ta oxide layers on plutonium and on HNO_3 concentrations could be observed, whereas a small decrease of this quantity would be seen from the experiments on Zr, if any change of DK was excluded (fig.2).

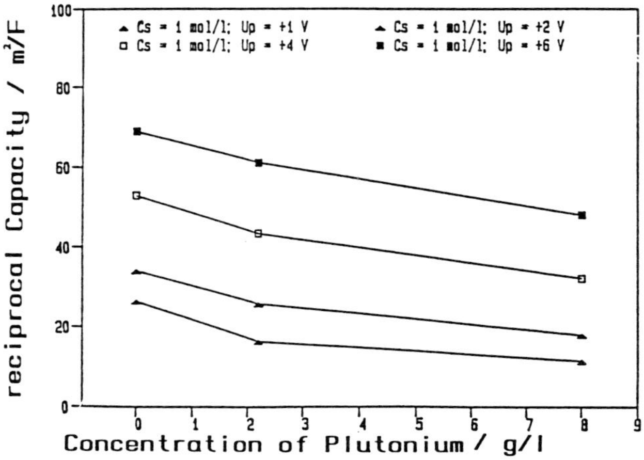

Fig. 2: Impedance on zirconium in 1M HNO_3 containing plutonium

3.2 EXPERIMENTS ON HAFNIUM IN THE CATHODIC RANGE

In the cathodic range the potentiostatic experiments on hafnium were made at various temperatures (20°C, 40°C, 60°C), various potentials (-1.0 V, -0.5 V, 0.0 V) and in various HNO_3 solutions (0.05, 0.5, 1.0, 3.0 mol/l). All the annual corrosion rates obtained under the various conditions are within the range of 10-5 to 10-3 mm/a. The values for the special case at -0.5 V are shown in fig.(3). These small corrosion rates can only be detected by RIM due to its high sensitivity. They would have always been neglected, if only electrochemical methods had been applied. Performing potentiodynamic measurements the electrodes were prepolarized up to +2 V in order to guarantee identical starting positions, this procedure being followed by cathodic experiments.

Fig. (4) shows that at 20°C and in HNO_3 systems from 0.5 to 5 mol/l, the oxide layer formed at +2 V is not dissolved even by a cathodic polarization at -1 V for 20h. Only 3M HNO_3 dissolves it. At higher temperatures on the other hand the oxide layer previously formed is always destroyed by cathodic polarization [3,10].

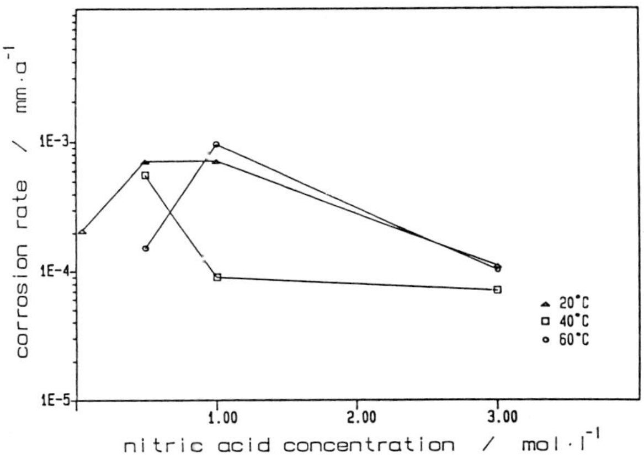

Fig. 3: Corrosion rates of hafnium in pure HNO_3 at -0.5 V

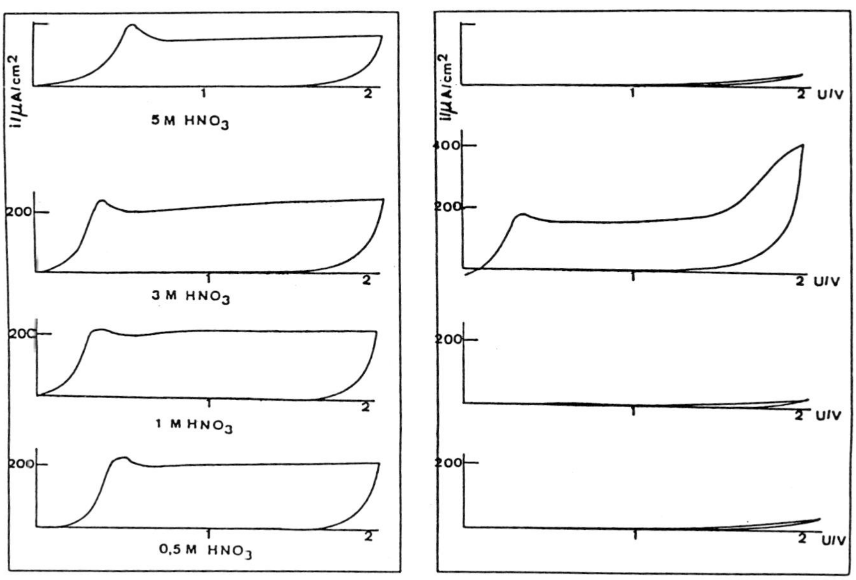

Fig. 4: Anodic polarization before (left) and after (right) a cathodic longterm experiment at -1V (20 h; 20°C)

3.3 CORROSION OF HAFNIUM UNDER THE INFLUENCE OF PLUTONIUM

In order to investigate the influence of plutonium on the corrosion rate of hafnium in the cathodic range experiments were made in 0.5, 1.0 and 3.0 M HNO_3 at various temperatures (20°C, 40°C, 60°C), and after prepolarization (+2 V), at potentials of (-1.0, -0.5, 0.0 V), the electrolyte solutions also containing 2.2 g/l Pu.

The results prove that plutonium does not influence the corrosion rate of Hf in HNO_3, these results being in accordance with those already obtained from the experiments performed in the anodic range. The amount of plutonium incorporated in the HfO_2 layer represents the practical conditions, since an external Pu influence on the oxide layer can be excluded due to the small range of the α-radiation in solution.

3.4 CORROSION OF HAFNIUM UNDER THE INFLUENCE OF URANIUM

For optimizing the Purex process also the influence of uranium on the corrosion of hafnium had to be examined. Therefore Hf corrosion rates at 20°C, 40°C, 60°C and in 3M HNO_3 to, which 20 and 60 g/l uranium was added, were investigated. In these systems the Hf corrosion rates increases with increasing temperature (fig.5), the values being irrelevant from the practical point of view. Even a small stabilizing effect of the uranium towards corrosion can not be overlooked.

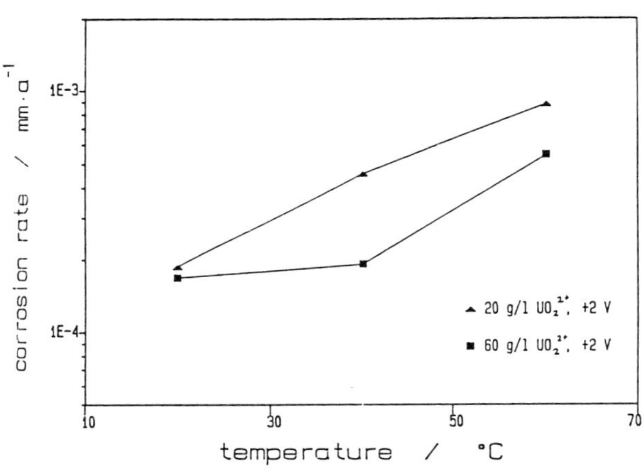

Fig. 5: Corrosion rates of hafnium in 3M HNO_3 + $UO_2(NO_3)_2$

3.5 THE INFLUENCE OF HALIDES ON THE CORROSION OF HAFNIUM

The corrosion rate of hafnium under the influence of F- for positive potentials at 20°C was already discussed in detail in [5,7,10] and it was shown that Zr diminishes any effect of F- by forming the complex ZrF^{3+}. In fig.6 the referring results obtained at various temperatures can be seen proving the Hf corrosion rate to be significantly higher than that, caused by the special parameters aforementioned. The rate increases definitely with incresing temperature and F- content. In practice the Zr content of the solution produced by the cladding is sufficiently high to mask any F- influence.

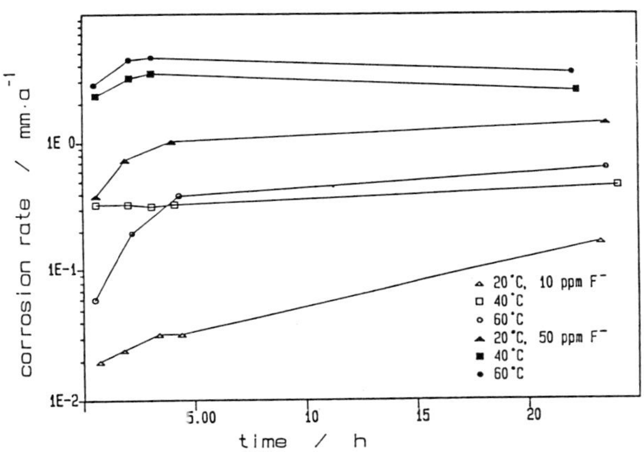

Fig. 6: Corrosion of hafnium under the influence of fluoride ions

Referring to the influence of Cl- it must be mentioned that in 0.1 - 0.5 M HCl solution the corrosion rates equal those in pure HNO_3, but increase drastically with higher Cl- content (fig.7).

3.6 CORROSION OF HAFNIUM AT REST POTENTIALS

At rest potentials the corrosion of hafnium was also examined at 20°C, 40°C, 60°C and in the HNO_3 concentration range from 0.1 to 8.0 mol/l. As fig.8 shows the potentials increase with rising HNO_3 concentration, this behaviour being due to the increasing negative partial current of HNO_3.

The corresponting corrosion rates measured lie within the m/a range. After passing a maximum the corrosion rates decrease with higher HNO_3

concentrations due to two opposing effects, the decreasing dissociation of HNO_3 and its increasing oxidation power.

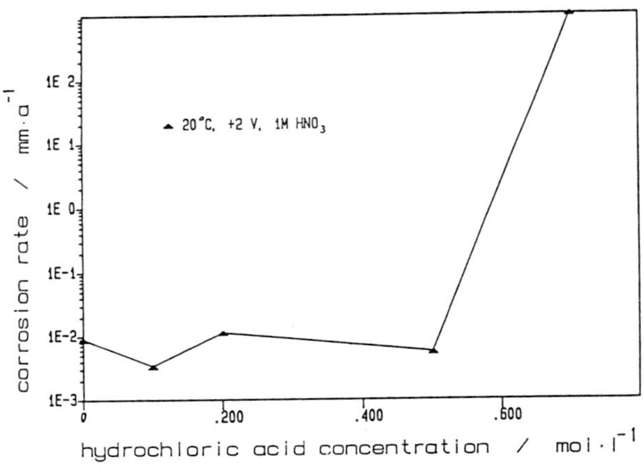

Fig. 7: Corrosion of hafnium under the influence of chloride ions

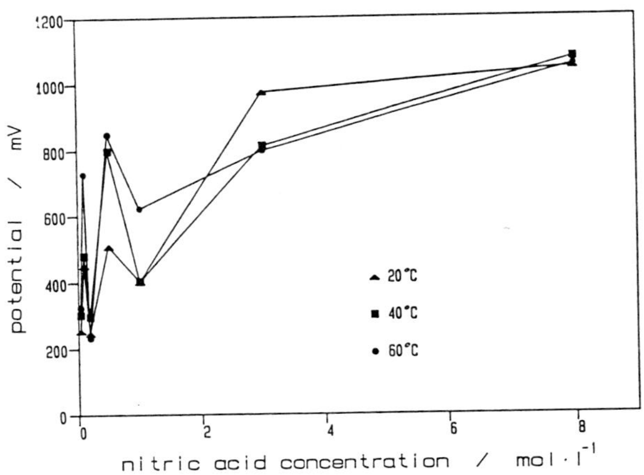

Fig. 8: Rest potentials of hafnium vs. nitric acid concentration

4.0 CORROSION OF TITANIUM

4.1 ANODIC POTENTIOSTATIC MEASUREMENTS

Anodic potentiostatic measurements on titanium were made at various potentials and at 20°C - 60°C, not only in pure nitric acid but also in HNO_3 (0.5 - 8 mol/l) containing 2.2 g/l Pu.

The corrosion rates are small and range from 1 μm/a to 12 μm/a (fig.10;11) and do not show any remarkable dependence on the Pu content and the potentials applied. Only a slight increase with temperature can be detected for pure HNO_3 systems.

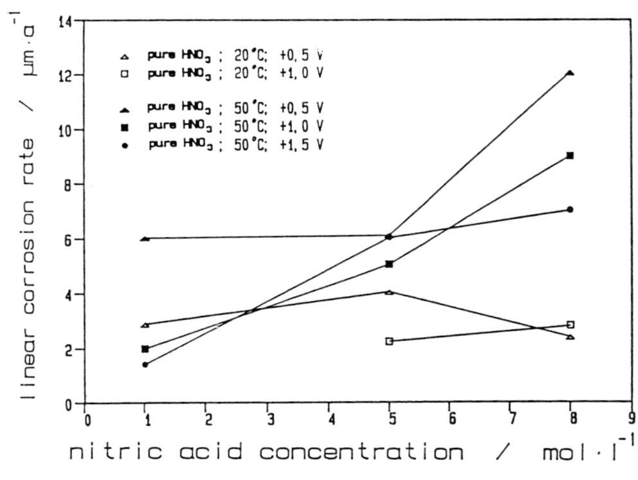

Fig. 9: Corrosion rates of titanium at anodic potentials

F- content (25 - 75 g/l) on the other hand drastically increases the corrosion up to 500 μm/a (fig.10).

4.2 CATHODIC POTENTIOSTATIC MEASUREMENTS

In the cathodic range experiments were performed at -0.2 V, -0.5 V and at 20°C in 1.0 - 8.0 M HNO_3 with pre - and unpolarized Ti electrodes (prepolarization at +2 V), the Pu content being 2.2 g/l or 4.4 g/l, respectively (fig. 11).

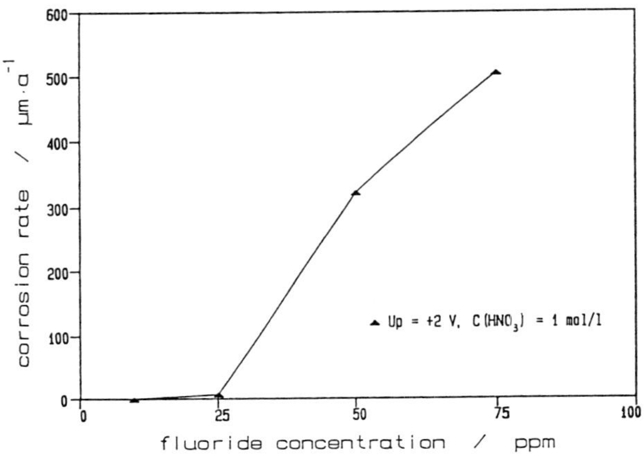

Fig. 10: Corrosion rates of titanium influenced by fluoride ions (20°C)

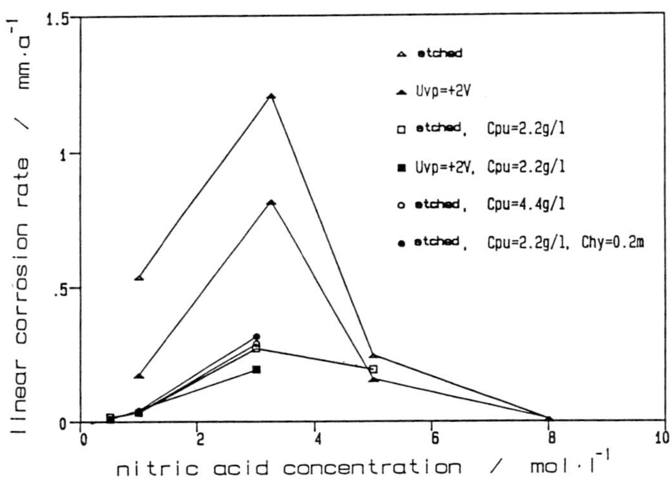

Fig. 11: Corrosion rates of titanium at -0.2V (20°C)

A maximum of the corrosion rate was registered in the 3 M HNO_3 system underlining the importance of this special concentration, which is due to the change of the dissociation characteristics of HNO_3 [10]. The decrease of the corrosion rate from 1 mm/a to 300 µm/a caused by the addition of Pu is obvious, whereas any further increase of Pu from 2.2 g/l to 4.4 g/l does not show any effect at all.

At -0.5 V there is an increase of the corrosion rate with HNO3 concentration into the mm/a range. All the TiO_2 layers produced by prepolarization at +2 V vanish in the cathodic range but the effect of that prepolarization results in a smoothing of the electrode surface which leads to smaller corrosion rates (0.9 mm/a instead of 1.3 mm/a).

4.3 CORROSION OF TITANIUM AT REST POTENTIALS

Titanium was examined at rest potentials in 1.0, 3.0, 5.0 M HNO_3 at 20°C and 40°C and in the presence of Pu (2.2 - 10 g/l). After 150 h the rest potentials were found to exist in the range of +0.7 to +0.9 V increasing with rising temperature, the corrosion rates ranging from 0.1 to 0.3 μm/a.

5.0 CORROSION OF ZIRCONIUM

5.1 ANODIC POTENTIOSTATIC MEASUREMENTS

The corrosion of zirconium was investigated by use of long term experiments (20 h) in 0.5 - 5 M HNO_3 at various temperatures (20 - 60°C) and at +2 V, but also under the influence of F^-.

The results show that Zr corrodes significantly even in solutions containing F^- in the ppm range, this effect being increased by rising temperature, which may be due to changes of the equilibria of the fluoride complexes (fig.12). The potential of local corrosion depends on the content of the electrolyte solution present, any increase of F- shifting the potential to more negative values. Pure HNO_3 without any F- on the other hand prevents any local corrosion up to +2 V. At higher concentrations it increases the stability of the oxide layer toward corrosion.

5.2 CATHODIC POTENTIOSTATIC MEASUREMENTS

The corrosion rates of Zr at -0.5 and -1.0 V at various temperatures (20°C, 40°C, 60°C) and HNO_3 concentrations (0.5 - 5 mol/l) are within the μm/a range. At higher temperatures 5 M HNO_3 causes a higher rate $\approx$ 1 mm/a the maximum being at 40°C (fig. 13). All corrosion

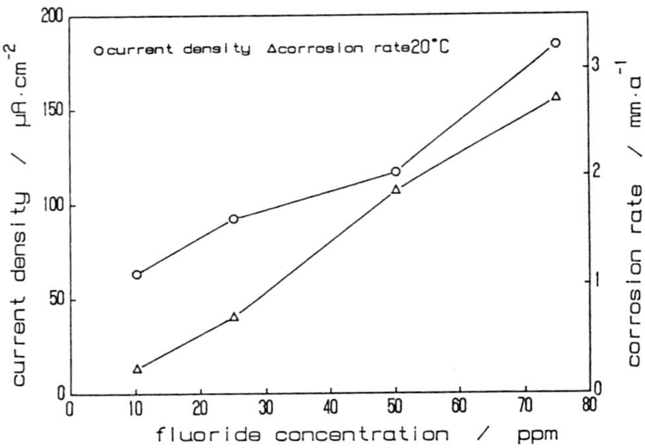

Fig. 12: Corrosion of zirconium influenced by fluoride ions

rates of Zr at -0.5 V and 20 - 60°C in 0.5 mol/l HNO_3 with 2.2 g/l Pu were below the detection limit.

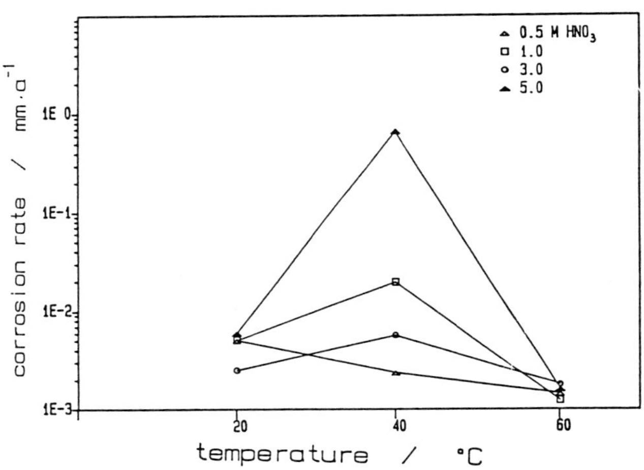

Fig. 13: Corrosion rate of zirconium at -0.5V vs. temperature

5.3 CORROSION OF ZIRCONIUM AT REST POTENTIAL UNDER THE INFLUENCE OF PLUTONIUM

Measurements at rest potential in 5 M HNO_3, containing 30 g/l Pu at 60°C show corrosion rates below the detection limit. These results prove that the corrosion rate of zirconium is also drastically reduced by the incorporation of Pu.

6.0 CORROSION OF TANTALUM

Tantalum is highly resistent towards corrosion in all ranges. Longterm anodic potentiostatic measurements at voltages from +1.0 to 10.0 V made in the HNO3 concentration range from 0.5 to 14.4 mol/l at temperatures from 20°C to 80°C resulted in corrosion rates below 5 nm/a. The same values were obtained from cathodic galvanostatic measurements in the same concentration range. At high current densities (10 mA/cm^2), on the other hand, hydrogen was developed which destroyed the electrode.

The investigations at rest potentials, also performed within HNO_3 concentrations from 0.5 to 14.4 mol/l resulted in corrosion rates near the detection limit, showing an increase of the rest potential with rising HNO_3 due to the rising negative partial current densities.

Anodic and cathodic measurements on tantalum were also made in presence of various ions (UO_2^{2+} , Ce^{4+}, Cl-, F-) proving that even high F- contents of ≈ 1000 ppm only lead to corrosion rates of 1 nm/a [10], the other ions having no influence at all.

So Ta will be the ideal material if only corrosion is taken into consideration.

7.0 CORROSION OF NIOBIUM

The low neutron cross section of Nb caused the corrosion of this metal to be investigated by use of Ta-182, which was contained in Nb in the 1000 ppm range, this procedure being justified on account of the chemical similarity of these two elements.

The annual corrosion rates of Nb were measured potentiostatically at +1 V and +2 V within the temperature range from 20°C to 60°C and the HNO_3 concentration range from 0.5 to 5.0 mol/l, as may be seen from fig.(14) the corrosion rates increase with rising HNO_3 concentration and temperature, the highest value being 800 nm/a at +2 V. Oxide layers produced by polarization will not be reduced when treated at - 75 mV, unless the temperature rises definitely higher than 20°C.

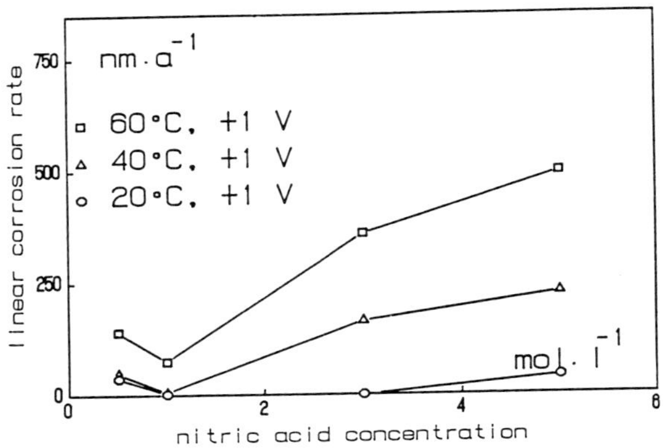

Fig. 14: Corrosion rates of niobium in HNO_3 at +1V

Also the rest potentials of Nb depend very much on temperature. The decrease of the rest potentials from 600 mV at 20°C to 450 mV at 60°C can only be understood if the increase of the anodic partial current with increasing temperature is taken into consideration. Evidence for this effect is provided by the rise of the corrosion rate for two orders of magnitude.

8.0 COMPARISON OF THE INVESTIGATED VALVE METALS

In the anodic range the annual corrosion rates of the investigated valve metals obtained from long term anodic potentiostatic measurements in HNO_3 systems (0.5 - 14 mol/l) at various temperatures (20 - 60°C) and potentials are comparatively small. For Hf they are in the range from 10 to 50 μm/a, for Zr from 1 to 10 μm/a at +2 V, but at +5 V even up to 20 μm/a and for Ti(1 - 12) μm/a. Ta and Nb are extremely resistent towards corrosion.

The corrosion rates determined under negative potentials are also very small. They range from 0.01 to 1 μm/a for Hf, and they are within the μm/a range for Zr, except those in 5M HNO_3, which climb up to 100 or even 1000 μm/a. For Ti the corrosion rates at negative potentials are near to some $\mu m/a$, increasing to 5000 $\mu m/a$ for 5M or 8M HNO_3. Also under these circumstances the corrosion rates of Ta are negligible. The rest potentials of all valve metals increase with rising temperature and HNO_3 concentration. The referring corrosion rates are within the $\mu m/a$ range, passing a small maximum with rising HNO_3 concentration, which may be due to the two opposing effects of HNO_3, the decreasing dissociation and increasing oxidation power.

In general plutonium increases the resistance towards corrosion, in all cases where the Pu concentration was 2.2 g/l, but also in some cases where the Pu concentration was significantly higher, up to 30 g/l. This result definitely contradicts all the suspected opinions made due to the higher LET of α-radiation. Further investigations will identify the real reasons, those experiments having already started with the detailed examination of the $(Hf;Pu)O_2$ layer.

There was no influence of $UO_2(NO_3)_2$ to be found on the corrosion of the investigated metals.

Fluoride-ions have a remarkable influence on the corrosion of the valve metals especially on Zr unless they are not masked by Zr as ZrF^{3+} complexes. Higher concentrations of Cl- ions act into the same direction. Especially Ta but to a minor grade also Nb resists corrosion even under the influence of halides.

From the practical point of view the best material resisting corrosion under Purex conditions is Ta, followed by Nb and also Hf. In some respect Ti and especially Zr seems not to be able to cope with the problems they are faced with in the field of corrosion under practical conditions of spent fuel processing.

ACKNOWLEDGMENTS

This work was carried out under contract to the Bundesministerium für Forschung und Technologie (Contract No. 02U 55047). Their financial assistance is gratefully acknowledged. Furthermore we would like to express our gratitude to Dipl.-Chem. J. Berger, J. Grewe and D. Wille for their practical help.

LITERATURE

[1] Baumgärtner, F.(ed.): Chemie der Nuklearen Entsorgung I+II; Verlag Karl Thiemig,München (1978)

[2] Schmieder, H. et al.: KfK-Bericht 3775 (1984) 181

[3] Erben, W.: Dissertation, Freie Universität Berlin 1985

[4] Droste, R.: Dissertation, Freie Universität Berlin 1987

[5] Marx,G., A. Bestanpouri, W. Erben, D. Wegen: Atomkernenergie Kerntechnik 48 (1986), 16

[6] Marx,G., A. Bestanpouri, C. Binder, W. Erben: Nukleare Entsorgung 1 (1981), 173

[7] Marx,G., A. Bestanpouri, R. Droste, W. Erben, W. Schönemann, D. Wegen:Dechema Vol. 102 (1986), 255

[8] Erben,W., A. Bestanpouri, G. Marx, J. Naegele, W. Müller: Radiochem. Measurements of Surface Corrosion Influenced by α-Emitters; 13ème Journées des Actnides, Eilat, Israel 1983

[9] Bjalobzeskjj, A.: Korrosion durch radioaktive Strahlung, Akademie Verlag, Berlin 1971

[10] Marx,G., A. Bestanpouri, R. Droste, W.Schönemann, D. Wegen: Final report BMFT - FK 02U 55047 (Bonn) 1987

[11] Bestanpouri,A., R. Droste, W. Erben, W. Schönemann, D. Wegen, G.Marx: Fresenius Z. Anal. Chem. 319 (1984), 848

Aerosol source term in the dissolver off-gas of a reprocessing plant

J. Furrer, A. Linek, F.J. Herrmann, B. Nemes

SUMMARY

The report describes measurements to determine the aerosol distribtion during simulated shear and dissolver operations in the test unit WAESCHE/PASSAT. These test rigs, an off-gas simulation facility and a prototyp off-gas filter system, serve to study the aerosol and iodine behavior and distribution as it will occur in the head-end of a reprocessing plant. Results of the aerosol production during simulated dissolution are compared with measurements, performed in the Karlsruhe Reprocessing Pland (WAK) in hot operation during several dissolutions of spent fuel elements.

J. Furrer, A. Linek, Kernforschungszentrum Karlsruhe GmbH,
Postfach 36 40, D-7500 Karlsruhe, Federal Republic of Germany
F. J. Herrmann, B. Nemes, Wiederaufbereitungsanlage Karlsruhe GmbH,
Postfach 1 20, D-7514 Eggenstein-Leopoldshafen, Federal Republic of Germany

1 INTRODUCTION

During the reprocessing of spent fuel elements two different off-gas streams in the PUREX-process occur:

1. the dissolver off-gas (DOG) and
2. the vessel off-gases (VOG) (Fig. 1).

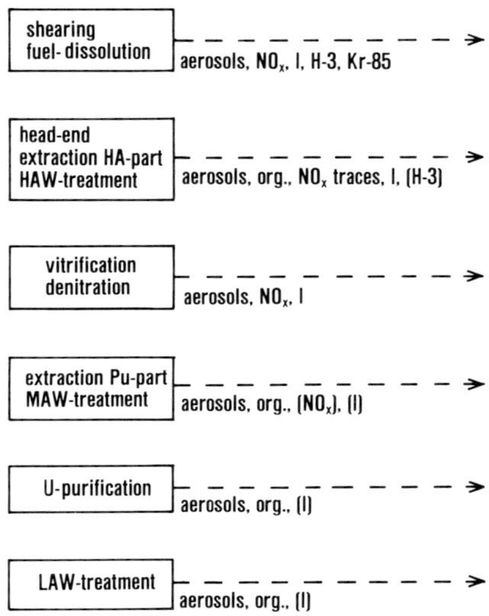

Fig. 1: Major process off-gas lines

The DOG shows a relatively low flow rate of 50 - 150 m^3/h and high fractions of radioactive materials. On the other hand the flow rates of the VOG amount to 500 - 5000 m^3/h with relatively low fractions of radioactive materials and high fractions of organic pollutants.

For testing a dissolver off-gas system the prototype filter system PASSAT and the dissolver off-gas simulation facility WAESCHE were coupled together (Fig. 2). WAESCHE consists of a dissolver, a condenser, and a NO_2-scrubber.

Evaluations of the aerosol source term and distribution in the dissolver off-gas simulation and treatment facility WAESCHE/PASSAT were done at the Kernforschungszentrum Karlsruhe (KfK) in

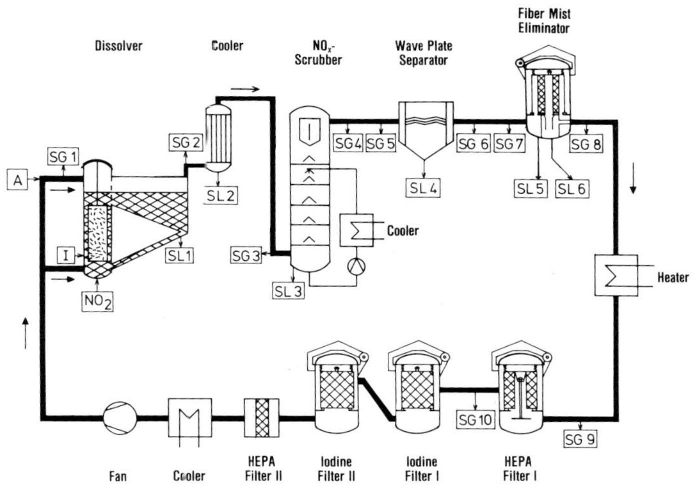

Fig. 2: Flowsheet of the dissolver off-gas simulation and treatment facility (WAESCHE-PASSAT)

nonradioactive operations. In the course of this test programme, investigations were made of the aerosol source term during the shearing and dissolution of spent fuel elements under simulated operational conditions.

For comparison with the inactive tests, samples were taken in the active dissolver off-gas during dissolution of spent fuel elements in cooperation with the Wiederaufarbeitungsanlage Karlsruhe (WAK).

The scope of the investigations in both facilities is to provide, to the design engineers and the involved licensing autorities, realistic data concerning the calculation of aerosol emissions and the optimal design of an off-gas cleaning system in the future German reprocessing plant /1,2/.

In WAESCHE an off-gas can be simulated containing droplets, shear particles, nitric oxides, iodine, and vapor. Fig. 2 shows the sampling points for gas (marked "SG") and for liquid streams (marked "SL").

2 BEHAVIOR OF SIMULATED SHEAR AEROSOLS IN THE TEST FACILITY WAESCHE/PASSAT

In the future reprocessing plant, one metric ton of spent fuel elements will be sheared and transferred into the dissolver basket during a period of one hour. An aerosol mass concentration of 50 - 250 mg/m^3 in the shear off-gas is assumed for a volumetric flow rate of 92 m^3/h.

The shear dust will be partially carried through the vapor chamber of the dissolver into the off-gas treatment section. The dissolver solution will have a temperature of 30 °C during shearing. Inside the NO_2-scrubber, a 1 molar nitric acid will be circulated at a similar temperature.

To simulate the shear dust in WAESCHE/PASSAT, lead monoxide (PbO) was selected. It resembles uranium dioxide (UO_2) in its physical and chemical properties.

Pulverized PbO was introduced into the upstream gas of the simulated dissolver. Samples were taken isokinetically upstream and downstream of the WAESCHE components, the wave plate separator, and the packed fibre mist eliminator. The evaluation of the PbO masses on the millipore filters was performed with an atom absorption spectrometer (AAS) and the particle sizes were determined with a scanning electron microscope (SEM).

Fig. 3 shows the diameters of the PbO-particles in the upstream gas of the dissolver ranged from 0.09 to 5.6 um. The maximum of the size distribution occurred for particles with a mean geometric diameter of 0.4 um. The aerosol concentration was determined to 210 ± 20 mg/m^3. Table 1 shows the results obtained during the tests performed.

Inside the dissolver, particles with a diameter ≥ 3 µm were removed with a DF of 3.7. Because of the low gas velocity of 0.16 m/s they settled out into the simulated dissolver solution. The condenser and the wave plate separator did not contribute to any retention. With the NO_2-scrubber and the packed fibre mist eliminator, a DF of 4.4 and 1800 was achieved. In 44 test runs, 73 % of the introduced PbO were removed in the dissolver, 21 % in the NO_2-scrubber, and about 6 % in the packed fibre mist eliminator.

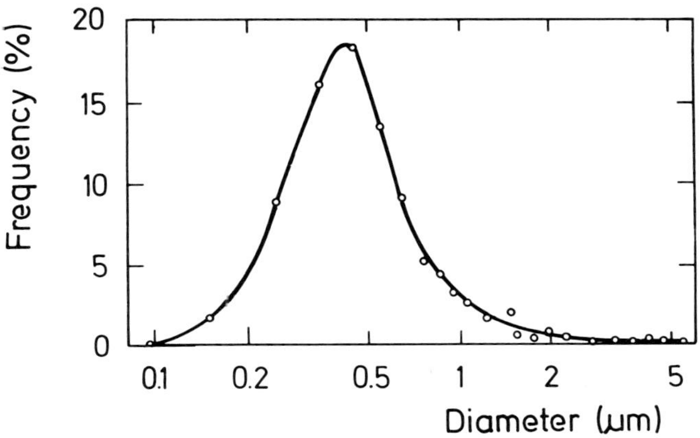

Fig. 3 Shear particles: PbO-particle size distribution (SEM interpretation)

Components	Dissolver	Condenser	NO_2-Scrubber	Wave Plate Separator	Fibre Mist Eliminator
DF	3.7 ± 0.4	1	4.4 ± 0.5	1	1800 ± 150
m(% PbO)	73		21		6

Table 1: Decontamination Factors (DF) for PbO-Particles and Removed PbO-Masses in WAESCHE/PASSAT during Simulated Shearing

3 BEHAVIOR OF AEROSOLS GENERATED IN THE WAESCHE-DISSOLVER

In the future German reprocessing plant the dissolver solution (4 molar nitric acid) will contain about 320 g Uranium/l (1.34 molar). The solution will be heated up to a temperature such that 10% of the dissolver inventory will evaporate per hour during the dissolution and iodine evaporation. The solution will be intermixed by stirring air (2.5 m_n^3/h). 20 m_n^3/h of nitric oxides will be generated. Droplets and steam will be transported to the condenser with 18 m_n^3/h leakage air. There the off-gas temperature will be reduced to 60 °C in order to condense large amounts of steam and simultaneously to prevent the condensation of elemental iodine. The condensate will flow back into the dissolver. In front of the NO_2-scrubber the off-gas flow rate will be increased to 150 m_n^3/h to maintain a constant-volume flow. The circulated liquid in the NO_2-scrubber will be nitric acid (1 molar). These conditions were chosen for the tests in WAESCHE. The only difference was the composition of the dissolver solution. Instead of an 1.34 molar U solution a 0.74 molar sodium nitrate ($NaNO_3$) solution, with 0.1 molar lithium nitrate ($LiNO_3$) as a tracer was used.

With WAESCHE, 43 dissolution test runs were simulated representing 5 h dissolution and 4 h iodine evaporation. Tab. 2 shows the DF and droplet flow rates in the downstream gas of some selected components of WAESCHE/PASSAT.

Component	Dissolver	Condenser	NO_2-Scrubber	Wave Plate Separator	Fibre Mist Eliminator
DF		98 ± 25	12 ± 4	1	590 ± 170
$\dot{V}_D$ (µl Solution/h	640	6.5	0.54		$0.1\ 10^{-3}$

Table 2: DF and Droplet Flow Rates (V_D) in the downstream gas of components in WAESCHE/PASSAT during simulated dissolution

The condenser, with a DF of 98, essentially contributes to return droplets to the dissolver inside the condenser. The droplet flow rate was reduced form 640 to 6.5 ul solution/h. In addition, the NO_2-scrubber and the fibre mist eliminator contribute to cleaning of the off-gas with a DF of 12 and 590, respectively.

4 AEROSOL SCOURCE TERMS AND RETENTION OF AEROSOLS IN THE DOG OF THE WAK

During 4 cycles in the shearing and dissolution of spent fuel elements, numerous samples were taken from upstream and downstream of the first DOG HEPA filter.

In Fig. 4, S_1 and S_2 show the sampling points. Parts of the off-gas were guided simultaneously through impactors, millipore filters and HEPA filters. The loaded impactors and millipore filters were used for evaluation of the particle sizes. For the determination of the radioactivity and isotope acitivity distribution, the particles collected of the HEPA filters were subjected to γ-spectrometry.

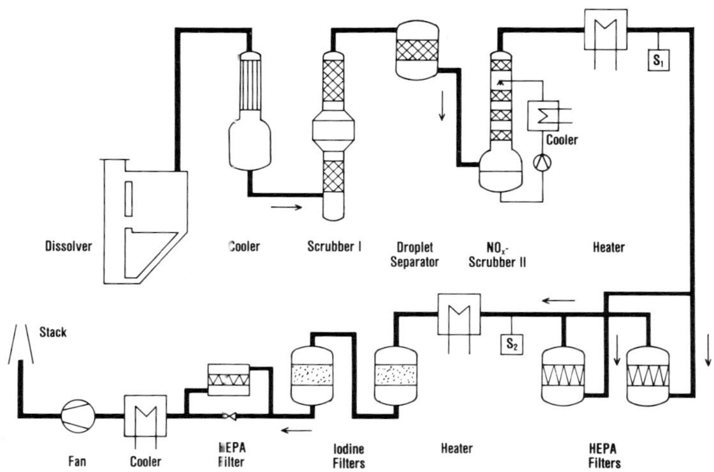

Fig. 4: Dissolver off-gas treatment in the WAK plant and aerosol sampling stations (S)

Fig. 5 shows the average mass concentration and the average radioactivity at the upstream side of the first HEPA filter during 4 shearing and dissolution processes. The mass concentration varied between 0.05 and 2.21 mg/m^3. The highest values were obtained at the beginning of boiling about 0.5 h after acid addition and reached a minimum during the iodine evaporation. The flow rate of the dissolver off-gas was 80 m^3/h.

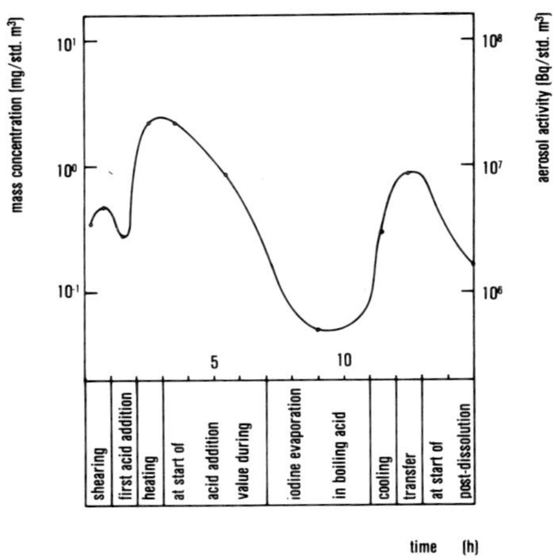

Fig. 5: Average values of the aerosol source term in the WAK dissolver off-gas upstream of the HEPA filter during a dissolution

The calculated decontamination factors of the first HEPA filter showed values > 10^5.

The particles collected on the impactors and millipore filters could not be used for determination of particle sizes. Both samplers were loaded with large masses of inactive aerosols partially caused by corrosion. The evaluation of the HEPA filters showed an isotope activity distribution similar to the feed solution. That means that no liquation took place during aerosol development. To estimate the aerosol source term in the dissolver off-gas, an overall decontamination factor of 20 was assumed for the components up to sampling point S_1 (Tab.3).

Considering this assumed overall DF the solid aerosol concentration can be calculated to 1 to 45 mg/m^3 on the downstream side of the WAK-dissolver.

At WAESCHE 0.64 ml/h of solution was carried out of the dissolver (see Tab.2). This correlates to a solid aerosol concentration of $22 \cdot 10^{-6}$ mol salt/m^3 /3/. Related on uranium nitrate (molecular weight:

502.18 g/mol), a concentration of 11 mg/m^3 can be calculated on the downstream side of the WAESCHE-sissolver. This value is comparable those measured on the downstream side of the WAK-dissolver.

Components in the WAK	DF
Cooler	2
1. Scrubber	5
Wire Mesh Droplet Separator	1
2. Scrubber	2
	20

Table 3: Assumed decontamination factors (DF) for aerosols of the components in the WAK

5 CONCLUSIONS

Investigations of aerosol behavior in dissolver off-gases were performed at WAESCHE/PASSAT and at the WAK.
Results:

1. The DF of the packed fibre mist eliminator in PASSAT was determined to be 1800 during tests with simulated shear aerosols and 590 during simulated dissolution in WAESCHE.
   Correspondingly high DF (>2000) were measured with the same type of fibre mist eliminator during the vitrification of highly radioactive waste in the PAMELA facility in Mol, Belgium /4/.

2. The HEPA filters both in PASSAT and in the WAK exhibited DF of >10^5.

3. During the simulated dissolution process in WAESCHE, the solid aerosol concentration, related to uranium nitrate, was calculated to 11 mg/m^3 carried out of the dissolver (gas flow rate: 20 m^3/h). An estimation for the dissolver off-gas of the WAK showed values from 1 to 45 mg/m^3 on the downstream side of the dissolver (gas flow rate: 80 m^3/h).

4. Aerosols removed behind the condenser cannot be carried back as liquid directly into the dissolver. Comparing the aerosol flow rate at that point, 0.65 µl solution/h ($\hat{\approx}$ 0.6 mg U/h) during simulated dissolution and 1.2...6.2 g PbO/h ($\hat{\approx}$ 1.3...6.6 g U/h) during simulated shearing were carried to the NO_2-scrubber. This means that aerosol flow rates behind the condenser during shearing are > 2000 times those during the simulated dissolution process.

In spite of the high aerosol charging of the dissolver off-gas during shearing of spent fuel elements, it is possible to avoid an enrichment of shear dust in the washing solution of the NO_2-scrubber and the flushing solution of the fibre mist eliminator. This objective could be accomplished with a shear durst separator installed parallel to the condenser (Fig.6). This separator would be in operation only during the shearing process. Afterwards the dissolver off-gas would again be passed through the condenser.

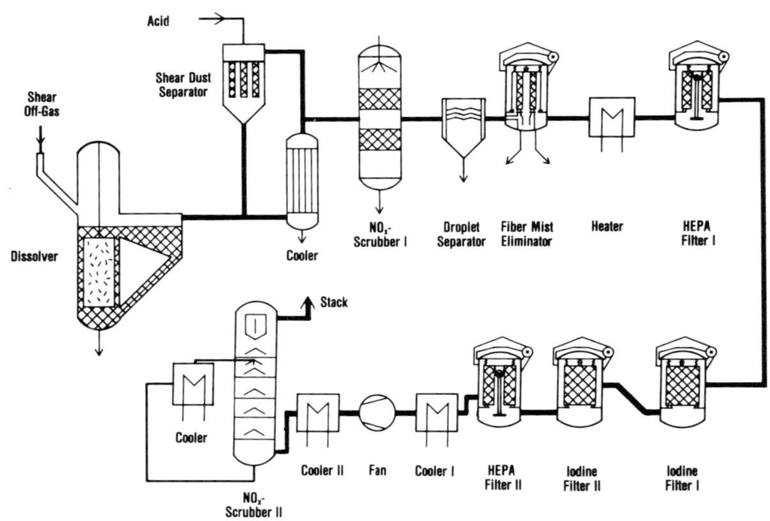

Fig. 6: Example of a flowsheet for future dissolver off-gas cleaning systems

6 LITERATURE

/1/ FURRER, J.; WILHELM J.G.; JANNAKOS, K.;
"Aerosol and Iodine Removal Systems for the Dissolver Off-Gas in a Large Fuel Reprocesing Plant".
CONF 780 819 (1979), pp. 494-511.

/2/ FURRER, J. et al.;
"Results of Cleaning Dissolver Off-Gas in the PASSAT Prototype Dissolver Off-Gas Filter System".
CONF 801 038 (1981), p. 566.

/3/ FURRER, J.; LINEK, A.; JANNAKOS, K.; HERMANN, F.J.; NEMES, B.;
"Aerosol-Source Term in the Head-End-Dissolver Off-Gas of a Reprocessing Plant".
CONF 860820, Vol. 2 (1987), p. 688.

/4/ HÖHLEIN, G.; WEISENBURGER, S.; TITTMANN, E.; WIESE, H.;
"Verglasung hochradioaktiver Spaltproduktlösungen in der Anlage PAMELA mit direkt beheiztem keramischen Schmelzofen."
KfK-Nachr. Jahrgang 18 2/86, S. 72-76.

Krypton removal from the dissolver off-gas with the solvent R-12

F. Weirich

SUMMARY

Compared to cryogenic distillation the essential characteristics of the cryogenic absorption process can be summarized as follows:

- low steady state Kr-85 inventory
  - low radiolytic solvent degradation
  - low shielding requirements
  - hot process demonstration possible with small amounts of Kr-85
- low sensivity to impurities
  - no preremoval of O_2, CO_2 or N_2O
  - compatible even with O_3 and CH_4
- cold box concept
- high decontamination factor
  high separation selectivity
  high product purity
- cryogenic absorption temperature
- operation at subambient pressure

No preremoval of O_2 is required. This is a considerable technical and safety advantage. If traces of O_3 are present in the off-gas they are separated together with the Xe. In the Xe raw product stream the O_3 can be decomposed to O_2 in a small CuO-catalyst bed.

The most important safety characteristics of the process are: Low Kr inventory, operation at sub-ambient pressure and no necessity to handle large amounts of H_2 for the removal of O_2. Moreover, the process is compatible with traces of methane and ozone in the off-gas. The low Kr inventory allows a hot demonstration of the process on full scale without a reprocessing plant. Such a hot demonstration prior to real application would be a considerable advantage. The state of development of our facilities and the experience gained with them is sufficient for the design of a demonstration plant.

F. Weirich, Kernforschungszentrum Karlsruhe, Institut für Heisse Chemie, Postfach 36 40, D-7500 Karlsruhe 1, Federal Republic of Germany

1 INTRODUCTION

During the dissolution of spent nuclear fuel the fisson product noble gases Kr and Xe are liberated into the dissolver off-gas. LWR fuel with a burnup of 36 GWd/t and 1 year cooling time contains about 1 m^3 inactive Xe and 0.1 m^3 fission product Kr per ton of uranium.

Only 6 % of the fisson product Kr is Kr-85 with an activity of about 10000 Ci/t and a half-life of 11a. The radiation exposure in the environment is only moderately reduced by a Kr-85 retention. The concentration of Kr-85 in the course of removal procedures creates a new hazard for the operating personnel.

Careful selection of a removal procedure from candidate processess should therefore guarantee, that the moderate radiation dose reduction for the environment does not occur at the expense of the plant operators. Process characteristics which take care of the safety of the operators are:

1. Reduction of operating personnel by automatic operation and control and by reliable equipment reduces maintenance operations.
2. Minimization of the Kr-85 inventory reduces the concequences of an accidental release and the radiation exposure during normal operation.
3. The probability of an accidental Kr-85 release can be reduced by operation at slightly sub-ambient pressure.

A cryogenic absorption process was developed, which takes into account especially these aspects.

2 BRIEF PROCESS DESCRIPTION

The cryogenic absorption process is a conventional gas absorption-desorption procedure at rather low temperatures and sligtly sub-ambient pressure. The basis of the process is the good solubility of rare gases in refrigerant-12 (dichlorodifluoromethane CCl_2F_2) and the poor solubility of air /1-7/. The distribution coefficients of the various off-gas components are shown in fig. 1.

Refrigerant-12 (tradenames: Frigen 12, Kaltron 12, Freon 12) is the usual refrigerant of home refrigeration units; it is non-toxic, inflammable and cheap.

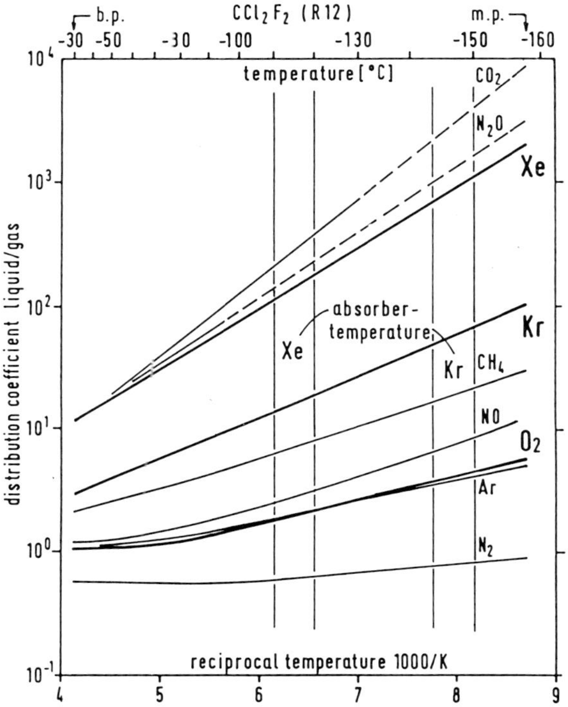

Fig. 1: Solubility of gases in liquid R-12

A simplified process flowsheet is shown in fig. 2.

Xe and Kr are removed in two successive absorption columns. The first column removes the Xe and the second column the less soluble Kr. In the intermediate stripper section the coabsorbed carrier gases are flushed back to the feed gas, to achieve a better product purity. The product gas is completely removed in the final stripper section. A pump recycles the unloaded solvent to the top of the absorber via a heat exchanger. The counter-current heat exchange between the cold and warm section of the solvent loop is needed to reduce the refrigeration energy requirements. The adsorber bed in the solvent loop serves to maintain a dry and clean solvent even in case of upstream malfunction.

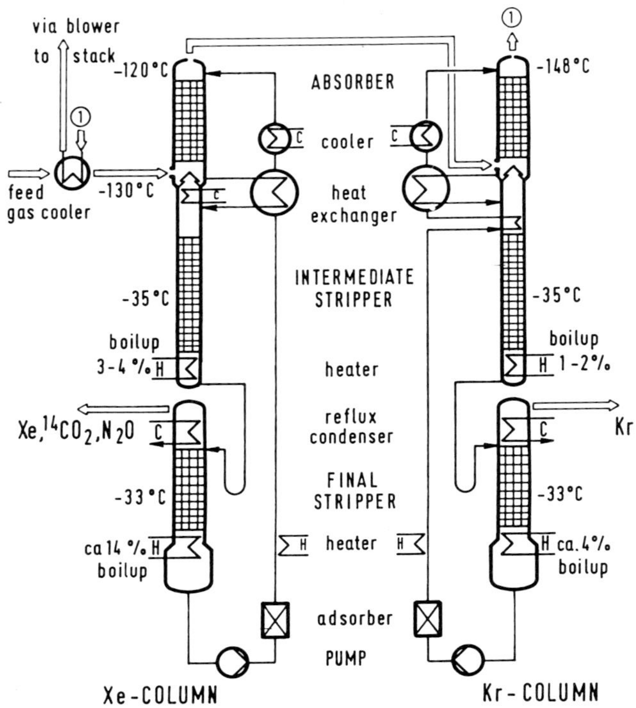

Fig. 2: Cryogenic absorption of rare gases (simplified process flowsheet at 0.8 bar)

2.1 FEED GAS COOLER

In case of malfunction of prepurification steps the feed gas may contain freezeable impurtities, especially water vapor. These can be removed in the feedgas cooler at about -130°C. At this temperature the small amounts of N_2O and C-14 as CO_2 in the dissolver off-gas are not removed, but separated lateron together with the Xe.

2.2 XE-ABSORBER

In the Xe-absorber Xe, N_2O and CO_2 are selectively removed by a counter-current flow of cold solvent. A small fraction of air and Kr is coabsorbed. From the Xe-free off-gas the Kr is absorbed in the subsequent Kr-absorber.

2.3 SOLVENT HEAT EXCHANGER

The cold loaded solvent from the absorber is routed to a plate type heat exchanger for counter-current heat exchange with the warmer solvent stream from the solvent pump. This reduces the energy consumption of the process. The incomplete heat exchange is compensated for by small heaters and coolers in the corresponding streams. During warming up of the solvent to nearly boiling point temperature, most of the coabsorbed gas components are liberated. The remainder is completely liberated in the intermediate stripper section.

2.4 INTERMEDIATE STRIPPER

The purpose of the intermediate stripper section is to guarantee a high product purity. A sufficiently Kr-free Xe product is of special interest for commercial utilization. For this purpose, the Kr/Xe separation factor should be about 10^8. To achieve an efficient Kr separation, the boilup rate in the fractionator sump must be at least 4 % of the solvent flow rate. The R12 vapor rising counter-currently to the solvent in the packing strips the coabsorbed components and the Kr back into the feedgas. The solvent now contains only Xe, N_2O and CO_2 (possibly traces of O_3) and is routed to the final stripper via a syphon.

2.5 FINAL STRIPPER

The dissolved gas components Xe, N_2O and CO_2 are completely boiled off in the final stripper. A boilup rate of about 14 % of the solvent flow rate is sufficient. In the reflux condenser, the R12 vapor is removed from the product gases. To prevent freeze-out of N_2O or CO_2, the temperature in the reflux condenser should be higher then -80° C.

2.6 SOLVENT VESSEL AND CIRCULATION PUMP

The unloaded solvent from the final stripper is collected in the solvent vessel and the pump recycles it via the solvent heat exchanger and a controlled solvent cooler to the top of the absorber packing.

2.7 KR-COLUMN

The design of the Kr column is similar to the Xe column. The essential difference is that the heat exchanger in the solvent loop is divided into an internal and an external part. The internal part is the top of the fractionator packing. This is necessary for the Kr column only and has the following advantages:

1. Only a small amount of R12 vapor is carried into the cold solvent in the absorber. This reduces the warmup and results in a better heat exchange.
2. Less Kr is recycled to the absorber, resulting in a lower steady state Kr inventory. The undesirable radiolytic solvent degradation is reduced in proportion to the inventory.
3. Higher Kr inventories can be accumulated without break-through into the purified off-gas. This is of advantage, if the product take-off is interrupted for a longer time, e. g. in case of malfunction. The Kr can be accumulated without process interruption until suitable operating conditions are rearranged. This is a simple meeasure to control occasional malfunction.

2.8 COLUMN PACKING

The absorber, intermediate and final stripper sections contain an efficient wire mesh packing. In our pilot facilities, two different types of packing have been compared experimentally. The Xe column contains "Goodloe" packing from Metex company, USA and the Kr column contains "CY" packing from Sulzer company, Switzerland. Both packings are about equally efficient according to manufacturers data. The Goodloe packing takes about 8 and the Sulzer packing only 4 % of the column volume; the latter packing has the advantage of a somewhat higher throughput. Table 1 /8/ summarizes the essential design data of our pilot plant facilities.

3 PROCESS CONTROL SYSTEM

Due to the minimum of available personnel, the purchase of a control system has been considered absolutely nesessary for shift operation. The system should allow automatically controlled operation, collection and storage of data and should also simplify evaluation. Evaluation should be possible during operation as well as after having finished a series of experiments.

	Xe COLUMN		Kr COLUMN	
packing type (SS)	Metex "Goodloe"		Sulzer "CY"	
	diameter	height	diameter	height
absorber	0.1 m	1.8 m	0.1 m	3.0 m
internal heat exchanger	——		0.18 m square,	0.6 m
intermediate stripper	0.075 m	2.7 m	0.1 m	2.7 m
final stripper	0.1 m	1.8 m	0.1 m	1.9 m
total height		12 m		12 m
solvent rate	≤ 450 kg/h		≤ 900 kg/h	
gas rate	≤ 30 m^3 (STP)/h		≤ 25 m^3 (STP)/h	

Table 1: Design data of the cryogenic absorption pilot plant

The hardware used in the system in SIEMENS-SICOMP hardware. the software has been designed by K. JAKOB Company, Karlsruhe. The system contains two intelligent hardware components: a Front-End-Computer (FEC) installed directly at the process and a superimposed Personal Computer (PC). The FEC consists of members of the SICOMP MMC-216 family. Process signal formers of the SIMATIC series create the interface to the process. The FEC serves the process, i. e. it polls the interfaces and converts incoming data to physical values, controls the process etc. The measured values are either transferred on a parallel or serial line to the superimposed PC.

System handling happens on the PC, which is supplied with two monitors, a keyboard and an adequately dimensioned hard disk. One monitor displays the menues required to use and manipulate the system. On the second one, process- and curve-diagrams are displayed in high-resolution graphics. Parameters required by any feature of the system can be set even when the process is running. the applied multitasking operating system allows a mathematical evaluation during operation.

The control system is able to handle up to 512 inputs at a time. Devices equipped with a parallel or serial port can be connected to special components of the FEC and evaluated on-line.

Measured values can either be plotted on a DIN A3-Plotter or printed on a line printer /9/.

4 EXPERIMENTAL RESULTS AND OPERATING EXPERIENCES

More then 2000 **hours** of continuous operation with the pilot plant have been accumulated. Additional about 2000 hours were needed to characterize the various components. More then 100 different operating conditions have been characterized. After changing the operating conditions, new steady state conditions are attained in less then 1 hour /10/. These last process characteristics are connected with the rather low residence time of Xe and Kr in the columns. The range of suitable operating conditions is relatively broad; this eases operation and automatic control.

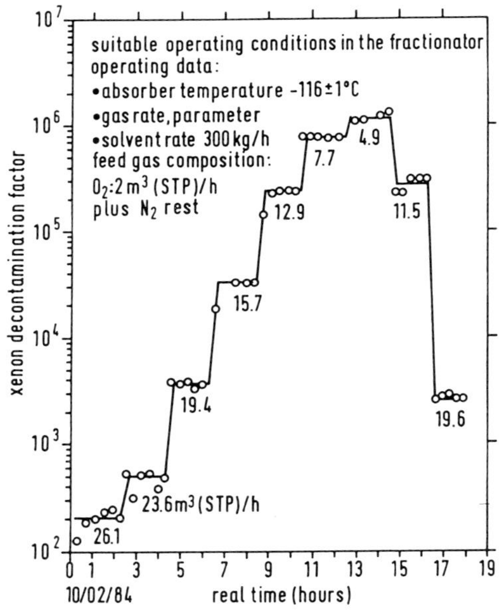

Fig. 3: Xe decontamination factors at stepwise changes of the gas rate

Figure 3 shows that relatively fast and large feedgas flow fluctuations can be tolerated without hydraulic problems. Short-time flow rate fluctuations of the circulating solvent of few percent do not cause trouble as shown in fig. 4. Also, short-time temperature fluctuations are smoothed out by the relatively large heat capacity of the equipment. The intermediate and final stripper sections are not in the main off-gas line and their operating conditions are independent of the feed gas flow. For an efficient operation of the intermediate stripper, the boilup rate in the sump and the temperature of the

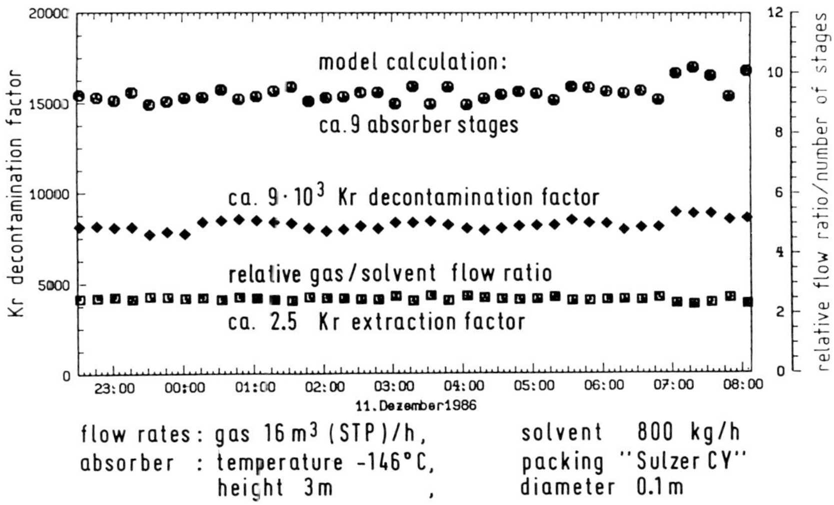

Fig. 4: Kryton column absorber operation

incoming solvent are related to each other. Suitable values have been determined experimentally and are consistent with model calculations.

A process model, based on stage theory has been developed and model predictions are in fair agreement with our experimental data /11/. A reference plant for 100 m^3/h off-gas troughput was designed, based on model predictions. The model proved to be reliable over a wide range of ooperating conditions checked in our pilot plant. The essential results of our pilot plant experiments are summarized in table 2.

investigated feed gas components: carrier gases N_2, O_2, Ar
product gases Xe, CO_2, N_2O; Kr impurities O_3, CH_4, NO

	Xe-column	Kr-column
absorber, temperature	-120 °C	-148 °C
decontamination factor	> 10^5	> 10^3
product purity (without R 12)	> 99.9 %	> 99.9 %
separation factor	≥ 10^6 Kr	> 10 CH_4/>10^7 O_2
product residence time	< 0.2 h	≤ 0.4 h
absorber stage height (HETP)	≤ 0.2 m	≤ 0.4 m

Table 2: Cryogenic absorption plant
(Summary of experimental results)

High decontamination factors and product purities have been attained. CO_2 and N_2O decontamination factors in the Xe-column are at least as high as the Xe-DF. The low Kr and Xe residence times show the low inventory obtained under optimum conditions. The extremley low Kr content in the Xe raw product deserves special attention and demonstrates the high separation efficiency of the process. A good separation of methane from Kr has also been achieved. Decontamination factors and product purities are not fixed values, but can be influenced at will by choosing suitable design and operation conditions. The essential data from the pilot plant experiments are the stage heights (HETP) for the packings. Per meter of packed height there are about 3 stages for Kr absorption and at least 5 stages for Xe absorption. At least 5 stages per meter have been found for all gas components under stripping conditions.

5 RADIOLYTIC SOLVENT DEGRADATION

Radiolytic solvent degradation has been investigated at the TU München /12/. From these data and our experimentally determined Kr residence time, the solvent degradation rate can be calculated. For a 350 tonnes per year plant the result is a rather low degradation of about 1 kg/a. Due to the small amount of potentially corrosive degradation products, corrosion is not expected to be a serious problem. Moreover, the solvent is kept dry with an adsorber bed and prepurification steps. A dry solvent is usually a sufficient protection against corrosion.

LITERATURE

/ 1/ E. Henrich, R. Hüfner, F. Weirich, W. Bumiller, A. Wolff; Proc. 18th DOE Nucl. Airborne Waste Management and Air Cleaning Conf., Baltimore 1984, p. 959

/ 2/ E. Henrich, R. Hüfner, F. Weirich; KfK-Nachrichten, 14, 172 (1982)

/ 3/ E. Henrich, R. Hüfner, F. Weirich, T. Fritsch; Removal of Xenon by Pressureless Absorption in Freon-12, in Radioactive Waste Management Series Vol. 10, (W. Hebel, G. Cottone, Eds.), Chur 1983

/ 4/ E. Henrich, R. von Ammon, K. Ebert; Krypton Control Alternatives, Proc. 18th DOE Nuclear Airborne Waste Management and Air Cleaning Conf., Baltimore 1984, p. 1004

/ 5/ E. Henich, R. von Ammon; Das Konzept der Abgasreinigung in Wiederaufarbeitungsanlagen, Atomenergie-Kerntechnik 46, 81 (1985)

/ 6/ E. Henrich, R. Hüfner; Modular Design of a Reprocessing Plant Dissolver Off-gas, IAEA-CN-43/447 (1983)

/ 7/ E. Henrich, U. Bauder, F. Weirich, T. Fritsch; Entwicklung integrierter Auflöserabgassysteme, Proc. Europ. Conf. Gaseous Effluent Treatment in Nuclear Installations (G. Fraser, F. Luykx, Eds.), Luxemburg 1985, p. 377

/ 8/ E. Henrich, U. Bauder, F. Weirich; Proc. Recod 87, Paris 1987, p. 563

/ 9/ K. Jakob, MSR-Technik, Karlsruhe, unpublished 1987

/10/ F. Weirich, E. Henrich; Jahrestagung Kerntechnik, Karlsruhe, FRG 1987, p. 397

/11/ E. Henrich, J. Röner, F. Weirich; unpublished 1985/86

/12/ W. Fürst, H. Heusinger; Atomenergie-Kerntechnik 45, 111 (1984)

Monitors for process control

H.-J. Becker

SUMMARY

It is shown how the PUREX process in an industrial reprocessing plant can be controlled by suitable monitors and what experience was gained during testing such monitors in the WAK reprocessing plant.

For the control of criticality a monitor provides a non-destructive determination of the criticality parameter k_{∞}^{max} by a combination of active and passive neutron measurements. By the same combination the residual fissile material content of high active waste solutions can be determined.

The detection of fissile material can also be performed by a simple passive neutron measurement, as shown by the results obtained with a passive neutron monitor installed at the dissolver of the WAK plant.

The on-line control of uranium and plutonium concentrations during extraction can either be done by a simple measurement of the temperature profile along the extraction apparatus or again by a passive neutron measurement.

X-ray-fluorescence is used as an additional method for determination of low residual uranium concentrations in waste streams.

Dr. H.-J. Becker, Wiederaufarbeitungsanlage Karlsruhe Betriebsgesellschaft mbH, Postfach 2 20, D-7514 Eggenstein-Leopoldshafen 2, Federal Republic of Germany

1 INTRODUCTION

In the reprocessing of spent fuel using the PUREX process monitors are widely used for process control. The principle of measurement, the task for the monitor, its construction and the location of installation change from plant to plant according to necessity and according to the state-of-the-art. Thus it may be that monitors which proved useful in the past, may become unnecessary by a change in process philosophy. For example, a monitor controlling the efficiency of homogeneous poisoning in the acid used for fuel dissolution will be super flous, when heterogeneous poisoning or a restricted geometry is used to maintain criticality safety. On the other hand technical progress in instrumentation may allow the use of a new monitor in certain circumstances if its detection limits have been improved. For these reasons a lecture of this kind can only give a picture of the actual state of the art, although some methods - e. g. neutron measurements - have been widely used for a long time once they have proved valuable and only technical details are improved.

In this paper several examples are given of how the PUREX process in an industrial reprocessing plant can be controlled by suitable monitors. Our own experience with such monitors in the WAK reprocessing plant is included.

Generally one can differentiate three tasks for process controlling monitors:

- control of criticality
- detection of fissile material
- in- and on-line control of concentrations;

the last group may be subdivided according to:

- what is to be measured
- surrounding conditions
  (physical state, matrix)
- indication of measurement
  (continuous display of trend or alarm system).

2 CONTROL OF CRITICALITY

The transfer of fuel or solutions containing fissile material from one part of the plant to another with different precautions to maintain criticality safety can be controlled by monitors.
Two examples:
In the storage pool of a reprocessing plant criticality safety of the stored fuel is maintained by the geometrical position of the single fuel elements and - if necessary - by neutron absorbing material. When the fuel is dissolved, the concentration of the fissile material is changed as well as the geometry and the physical state. The criteria for criticality safety in the dissolver are now restriction of geometry and use of heterogeneous poisoning (the basket is made of hafnium, which is a neutron absorber). These criteria differ from those for the storage pool. For an initial enrichment up to 3.5 % U-235 the dissolver is always safe due to geometry, even for fresh fuel. for higher initial enrichments than 3.5 % it has to be proven that the criticality parameter k_{∞}^{max} (which is defined for fuel rods in an optimal moderating lattice) does not exceed a certain level. This is achieved by control of the data file for the fuel element and - independently - by a spent fuel element monitor. With this monitor the multiplication factor M for the neutron flux of the fuel element is determined by a combination of active and passive neutron measurement. According to a calibration curve the value of k_{∞}^{max} is derived from the measurement of M. An example of such a measurement is shown in Figure 1. After measurements on more than 100 fuel elements of various types performed in collaboration with the "Institute for Neutron Physics and Reactor Technology" of the Karslruhe Nuclear Research Centre (KfK) with an experimental monitor had proved the principle of the monitor /1/, now an industrial prototype of a spent fuel element monitor is under construction.

In the first extraction column uranium and plutonium are separated from the fission products. The highly active waste solution is then transferred to the vitrification process, and so leaves the geometrically safe HA-column. To maintain criticality safety in the following process steps it is planned to perform a measurement of residual fissile material content in the HAW. The monitor should alarm at a concentratin of 2.5 g Pu/litre and works on the basis of an active neutron measurement. To prove the principle of this measurement an experiment is planned for this year in the WAK plant. In this experiment plutonium will be added to real HAW solution and measured stepwise.

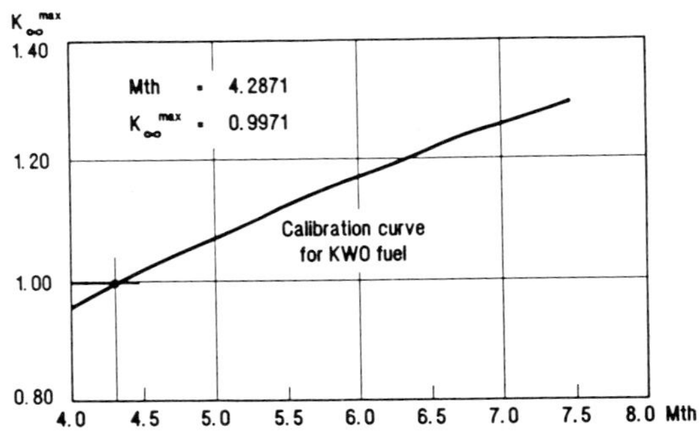

Fig. 1: Typical result of a measurement with a spent fuel element monitor

3 DETECTION OF FISSILE MATERIAL

For the detection of possible deposits of fissile material, which may occur over a longer period, passive neutron monitors must be provided on various process vessels. For demonstration of the feasibility of this technique under hot conditions passive neutron monitors were installed at the dissolver of the WAK plant. The monitors gained a high availability and provide direct information of the process in the dissolver, as shown in figure 2:

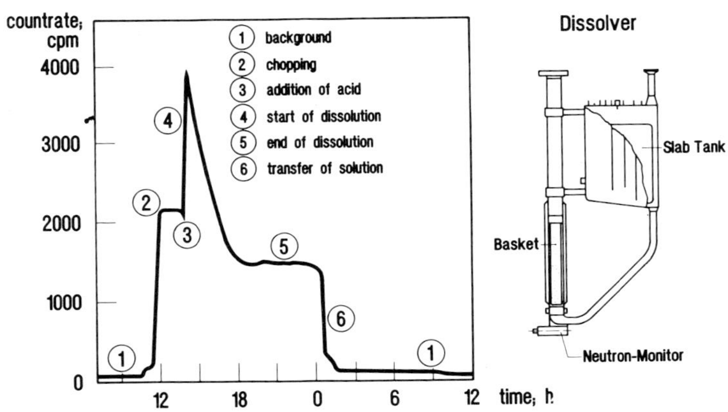

Fig. 2: Neutron count-rate during dissolution of fuel

When the fuel is chopped into the dissolver basket, the count-rate of the neutron monitor increases significantly above background level; addition of acid, start and end of the dissolution process itself can easily be recognized by the count-rate. When the dissolver solution is transferred, the countrate goes down to the background level again /2/. By observation of the background level countrate of the empty dissolver information on deposits of fissile material at the dissolver bottom can be gained. This would cause an increase in background count-rate. This effect was never observed in the several years of monitor operation.

4 ON-LINE-CONTROL OF CONCENTRATIONS

For process control purposes especially the control of the extraction process is important. This is shown in two examples:

In the HA-column uranium changes from the aqueous to the organic phase. Due to this reaction heat is generated. When you measure the temperature along the column, you will get the maximal temperature at that location where most of the uranium changes from the aqueous to the organic phase. This location is normally in the upper third of the column. If the extraction behavior of the column alters, this results also in an alteration in the profile of the temperature along the column, as shown in figure 3. The shift of the main extraction zone to the lower part of the column of the testrig UEZ is detected by the measurement of the temperature long before uranium (and plutonium) will appear in the HAW /3/.

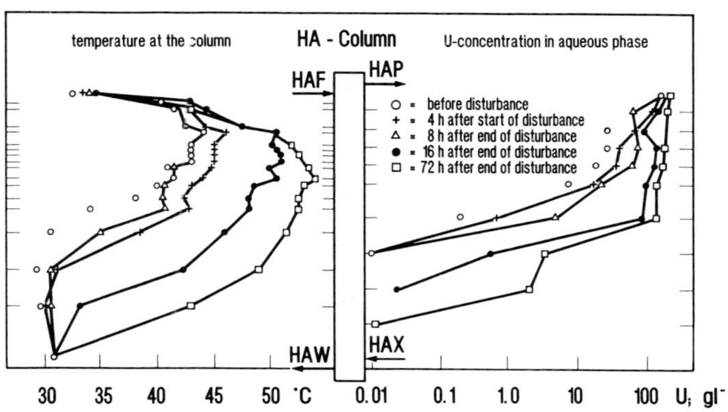

Fig. 3: Simulated malfunction in UEZ with deterioration of U-extraction

So an operator in a reprocessing plant can take countermeasures before the malfunction leads to solutions out of specification.

Just like the control of the separation of the fission products in the HA-column by a temperature measurement, the separation of U and Pu in the BX-column can be controlled by passive neutron monitors. Monitors installed at the surface of the column tube give on-line information of the Pu-concentration along the column as well as an information about their fluctuations in the course of time. This was proved during several campaigns in the plutonium test facility PUTE of KfK /4/. Figure 4 shows the arrangement of four monitors along the electrolytic BX-column ELKE. The decreasing Pu-concentration with increasing column height as derived from the measured count-rates at the moment is shown as well as its variation during the last hours.

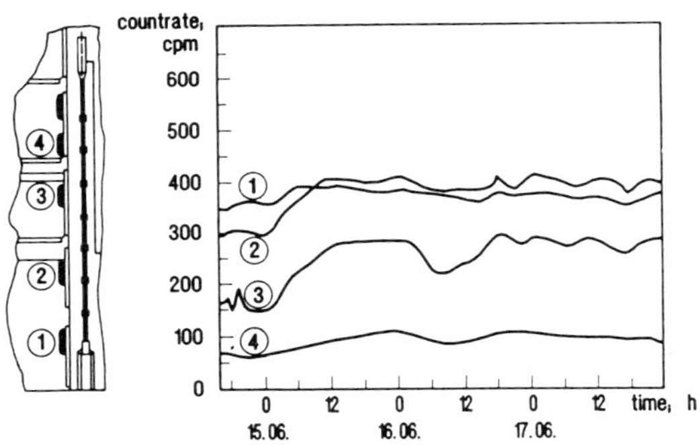

Fig. 4: Variation of Pu-profile in a pulse column

In a similar manner the extraction behavior of the electrochemically working BX mixer settler battery 1B-EMMA in the WAK plant is controlled. Neutron monitors installed on the top of the battery indicate the Pu-concentrations at the various stages of the battery (fig. 5). An initiating malfunction with the plutonium staying in the aqueous phase can be recognized early and countermeasures be taken.

In these two examples the monitors register the concentrations of the elements in the liquid phase (Pu-concentration directly; U-concentration indirectly). After draining and rinsing of the process vessel the neutron monitors provide further information on deposits of fissile material.

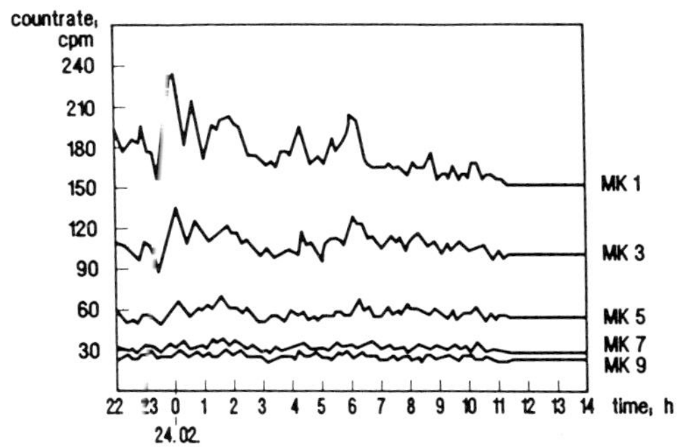

Fig. 5: Variation of the Pu-profile in the mixer settler battery 1B-EMMA

In other cases one is interested in detecting concentrations exceeding an operational level, as for example the concentration of residual uranium in 1CW. In the uranium extraction cycle UEZ of the TEKO facility an in-line instrument, which was developed by the "Institute for Hot Chemistry" of KfK, is used for this purpose. The encouraging results in the operation of this in-line monitor will be useful in the further development of this instrument, which are necessary before it comes into a "hot" use.

These examples show how monitors with their direct response which, in most cases, is nearly free of delay, can inform an operator of the actual situation in various parts of his plant and provide information about actual trends there. In this way he can recognize deviations from normal operation and can take countermeasures, which increase the safety in plant operation.

LITERATURE

/1/ H. Würz:
A Non-destructive Measurement System for Spent LWR Fuel Assemblies; Nuclear Safeguards Technology, Vol. II, IAEA-SM-260/30, 1983

/2/ H.-J. Becker, H. Würz:
Neutron Measurements at a Dissolver of a Reprocessing Plant; Int. Conf. on Analytical Chemistry in Nuclear Technology, Karslruhe, 3.-6.6.1985, Abstracts

/3/ K. Eiben, H.-R. Oeser:
Großkomponententest in der TEKO: Verfahrens- und Anlagenoptimierung für die Wiederaufarbeitungsanlage Wackersdorf, Jahrestagung Kerntechnik, Aachen, 8.-10.04.1986

/4/ H.-J. Becker, H. Würz:
Neutronenprofilmessungen an Pulskolonnen, 6. PWA-Statusbericht, Karlsruhe, 13.-14.3.1986, KfK-4177, 531-539

/5/ H.-J. Becker, H. Belzig, E. Gelfort, P. Groll, W. Herbert, K.R. Schulz:
Use of In-line Monitors in an Industrial Uranium Extraction Cycle, Int. Conf. on Analytical Chemistry in Nuclear Technology, Karlsruhe, 3.-6.6.1985, Abstracts

/6/ J. Furrer, H.-G. Dillmann:
Einrichtung zum Nachweis von Iodisotopen, Deutsches Patentamt Offenlegungsschrift DE 33 24523A1, Offenlegungstag 17.01.1985

Concept of an improved process control system in the uranium extraction cycle (PLURA)

D. Stritzke, H. Ritter

SUMMARY

In the experimental facility for testing components for reprocessing spent fuel elements (TEKO) it is planed to install an improved digital process control system in the 1. uranium extraction cycle (UEC). Operation of the UEC test and demonstration facility, and further advancement of control systems in chemical industry, have resulted in comprehensive control systems being added to the proven individual control concepts. In addition, the use of a digital process control system allows process control and plant management as well as safety and reliability of plant operation to be optimized. Ten functional units are being upgraded within the framework of the process control system and will be installed and operated in an interconnected system.

Dr. D. Stritzke, H. Ritter, Wiederaufarbeitungsanlage Karlsruhe Betriebsgesellschaft mbH, Postfach 2 20, 7514 Eggenstein-Leopoldshafen 2, Federal Republic of Germany

1 INTRODUCTION

An experimental facility for testing components for reprocessing spent fuel elements was built on the premises of the Karlsruhe Nuclear Research Center in 1975 - 1978. That facility (TEKO) is operated by WAK on behalf of DWK. The test rigs for component testing are set up in two halls of 600 and 400 m area, respectively. One hall contains the first uranium extraction cycle (UEC), including the necessary utilities, such as acid rectification, evaporators for reconcentrating the uranium solutions, and solvent purification.

The test runs conducted without plutonium were performed with depleted uranium and non-radioactive simulated fission products.

After each of the units had been studied individually for its hydraulic behavior under mass transfer conditions, and that behavior had been optimized, the units were run together for the first extraction cycle in accordance with the flowsheet conditions of the PUREX process. In this phase, detailed experience was accumulated in starting up and shutting down the plant and also during normal operation.

One important object of studies was the specific initiation of defects and breakdowns in operation and their repercussions on individual components and on the operation of the whole interconnected setup.

2 ADVANTAGES OF A DIGITAL PROCESS CONTROL SYSTEM

The technical measurement and control equipment of the UEC in its present version permits the whole system to be run in an interconnected mode only at high personnel expenditure. This is due to the fact that, when the facility was planned, attention was devoted especially to studies of single components and their optimization, as a consequence of which a large number of control actuating functions and readings must be made "on the spot". This means that additional sensors for the main process streams must be installed for an interconnected control of the whole UEC.

Operation of the UEC test and demonstration facility in the component testing hall, and the further advancement of control systems in chemical industry, have resulted in comprehensive control systems being added to the proven individual control concepts. However, this requires the reliable logging of plant operations data.

An interconnected control system offers the following advantages in use:

- The flows are kept largely constant with respect to their volumetric flow rates.
- Single fluctuations in the volumetric flow rates will not affect downstream functional units.
- Consequently, the product quality is kept constant.
- The availability of the plant in operation is enhanced.
- The use of modern microelectronics simplifies fault diagnosing.
- The higher degree of automation facilitates the work of the operating crews and avoids maloperation.

On the whole, this adds to the safety and reliability of plant operation.

In addition, the use of a digital process control system allows process control and plant management to be optimized in small control rooms. This is achieved by centralizing the display of process information and by process management by means of color video displays.

Bus data traffic, which saves line capacity, and software configuration and switching of the necessary functions will help to rationalize the sequences of planning, installation, and modification.

3 FUNCTIONAL UNITS IN THE UEC TO BE UPGRADED

The following functional units in the UEC are to be upgraded within the framework of the process control system to be newly installed and later to be operated in an interconnected system:

- HA column
- TBP steam stripper
- HS1 column
- HS2 mixer-settler
- BX column
- BXX chemicals makeup
- BS column

- C column
- CK mixer-settler
- uranium evaporator.

4 DESIGNING THE CONTROL SYSTEM FOR THE UEC

Figure 1 shows the structure of the PLURA process control system as a modern digital unit.

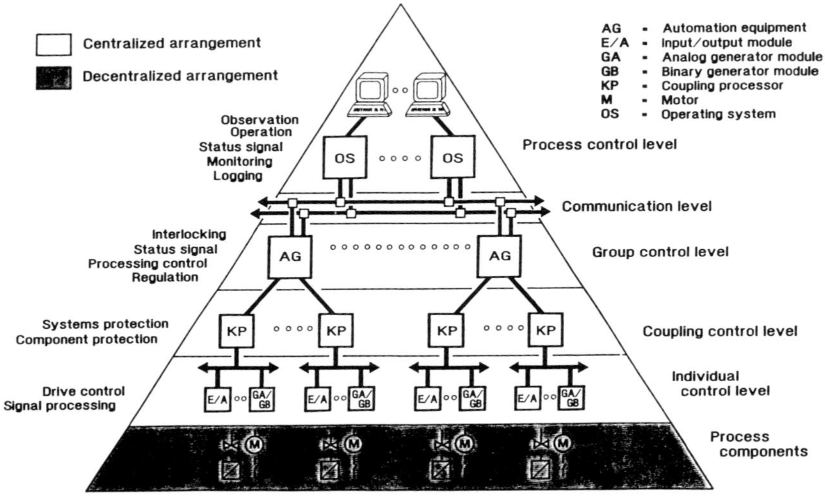

Fig. 1: Structure of the PLURA process control system

The individual levels provide information about the allocation of jobs.

Figure 2 shows the backup of the UEC control system by digital control technology.

The right side of the diagram indicates the existing, conventional control facility of the UEC designed in a parallel setup, i.e., a cable is run from each measuring point to the control panel.

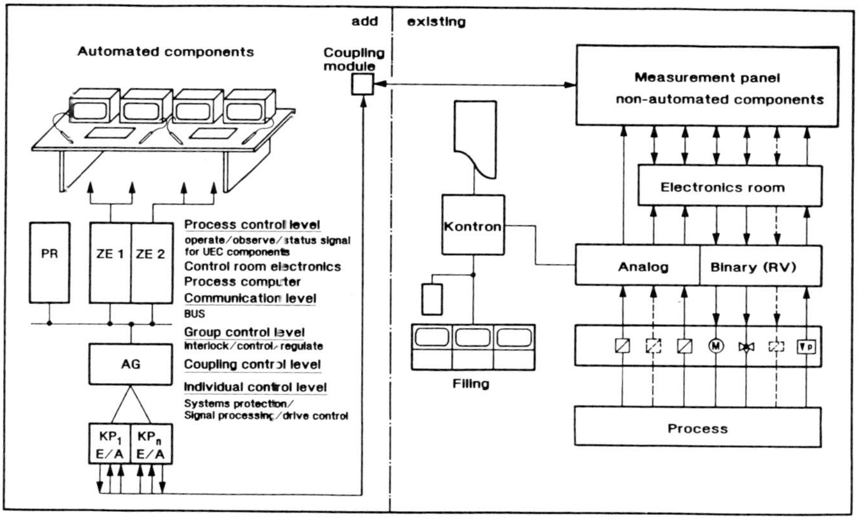

Fig. 2: Digital control systems supplementing the UEC control system

A modern filing system has already been backfitted and will assume the following functions:

- Long-time storage of measured data with graphic and alphanumeric outputs,
- long-time storage of analytical data with graphic and alpha-numeric outputs,
- long-time storage of logbook entries.

All data bases can have rapid access to specific data by means of search programs. These functions are preserved and, for this reason, are not duplicated in the new control system.

The left side of the diagram shows the levels of the new, planned control system. The new system will be coupled at a newly created interface in the control panel. The design corresponds to the

structure of the process control system shown above.

Figure 3 shows more details of the structure of the PLURA control system.

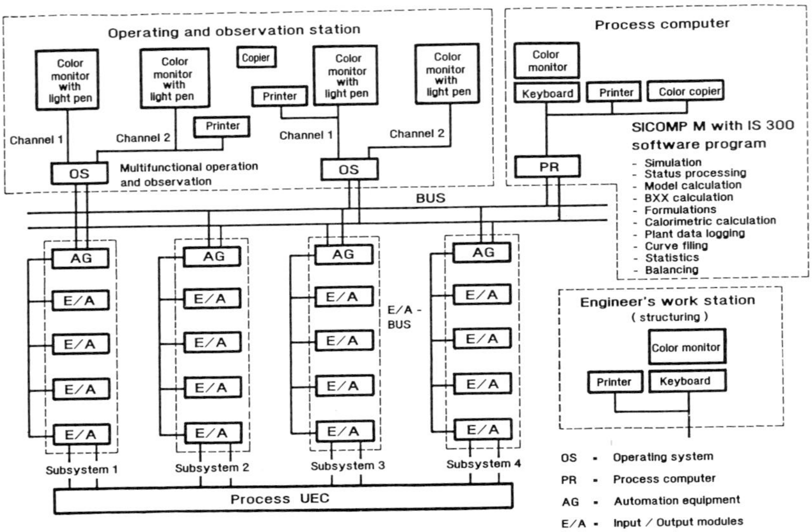

Fig. 3: The PLURA control system

The scope of the hardware and software is determined by the plant data:

Approximately 250 analog signals and 600 binary signals must be processed. More than 50 control circuits are required. The color displays will show some 30 graphs of the plant and 300 curves.

The availability of the plant, and the costs, require careful tailoring of the systems configuration. The equipment installed in the operations station and the bus are designed in an redundant fashion.

In the automation equipment of the group control level, the whole scope of the plant is distributed over a number of autonomous units, each of which is located in a cabinet with an independent power supply.

A separate work station for an engineer is provided for structuring the automation equipment. The control room also contains a process computer, which later is to perform the following jobs:

- process simulation
- calculations for formulations
- model observer
- calorimetric calculations
- process data logging
- balancing
- statistics.

The two systems of the operating station and the observation station contain the same scope of the program; consequently, all programs can be called up on each of the four screens. These are hierarchically structured programs for

- graphic displays of the plant flowsheets
- curve displays with six curves per image
- readouts (digital or as bar charts), e.g., of vessel filling levels
- statements in plain text.

5 EXPLANATION OF CONTROL STRATEGIES

The HAF volumetric flow rate is the reference variable of the whole process. The heavy metal throughput of the plant is controlled by variations of the volumetric flow rate. As it is shown by the equations in Fig. 4, all other volumetric flow rates are matched to the HAF volumetric flow rate. The differences in density, and the associated differences in volumetric flow rates between the flows entering and those leaving the extractors, which are caused by changes in loading of the phases, are described for the steady state of operation by the factors K_n (org.) and F_n (aqueous).

For instance, the factor $K_1 = V_{HAP}/V_{HAX} = 1.047$. This means that the organic volumetric flow rate at the top of the HA column is increased by 4.7% as a result of the uranium load. The control strategies will be explained below on the basis of the HA/HS1 column as an example.

The column is run in the aqueous disperse mode. The HAF volumetric flow, which contains HNO_3, is fed into the top of the column.

The organic solvent consisting of 30% TBP in n-dodecane is supplied to the bottom part of the column together with the recycled organic phase.

$$\dot{V}_{HAP} = \dot{V}_{HAF} \frac{P_{HA} P_{HS1} K_1}{P_{HS1} - P_{HA} K_1 F_2}$$

$$\dot{V}_{BXF} = \dot{V}_{HAF} P_{HA} P_{HS1} K_1 K_2 K_3 \left(\frac{1}{P_{HS1} - P_{HA} K_1 F_2} - \frac{P_{BS} K_5}{N} \right)$$

$$\dot{V}_{BSP} = \dot{V}_{HAF} \frac{P_{HA} P_{HS1} K_1 K_2 K_3 F_4 F_5}{N}$$

$$N = P_{HA} K_1 (P_{BS} K_5 F_2 F_4 - P_{BX} F_2) + P_{HS1} (P_{BX} - P_{BS} K_5 F_4)$$

P = o : a phase ratio

K_n, F_n = $f(\varrho, c, T)$

Fig. 4: Dependence of the product flows on HAF as a reference variable

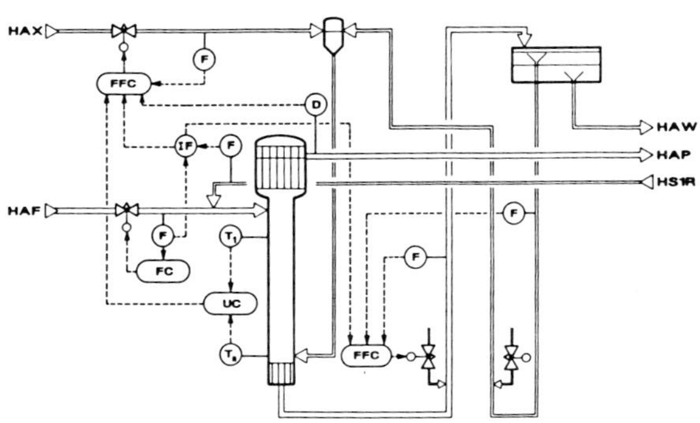

Fig. 5: The HA column of PLURA

At the base of the column, a mixed phase consisting of aqueous plus some organic phase is extracted by means of an airlift and supplied to a decanter above the column. This is where the two phases are separated. The aqueous phase, which contains the simulated fission product, is then treated further in the process.

At the top of the column, the organic phase with the extracted uranium is discharged and fed to the HS1 scrubbing column.

The aqueous reflux from the HS1 column is added to the HAF volumetric flow.

The phase ratio of the HA column is defined as follows and should be kept constant within a permissible bandwidth:

$$P_{HA} = \frac{\dot{V}_{HAF} + \dot{V}_{HS1R}}{\dot{V}_{HAX}} = \text{const.}$$

The control system described in Fig.4 is used to achieve this state.

The HAF volumetric flow rate is the reference variable for the whole process and is kept constant by fixed setpoint control.

The HS1R volumetric flow is added to the HAF volumetric flow, and this value is supplied to a ratio controller. In accordance with the preset phase ratio, the HAX solvent flow is set or matched.

The HAX volumetric flow rate is matched if there are fluctuations in the HS1R volumetric flow rate or if the extraction front in the column is shifted. The position of the extraction front in the HA column is indicated by the axial temperature profile because of the exothermal extraction process.

If, e.g., the extraction front moves downward, i.e., if more uranium would be discharged with the HAW than is permitted, the phase ratio is

changed by an increase in the HAX flow. A potentially faster indicator showing changes in the column is the density change in the HAP flow.

Both disturbance injections for ratio control, i.e.,

- changes in the temperature profile and
- density changes in the HAP,

will be tested.

5.1 MIXED-PHASE CONTROL

In order to ensure that no interface is formed in the bottom part of the column, a mixed phase is extracted which consists of the aqueous phase fraction plus some organic phase.

For this purpose, the combined signal of the HAF + HS1R volumetric flow rates is assigned a factor and supplied to the ratio controller for the mixed phase control.

On the other hand, however, a minimum quantity of organic phase must be transported to the decanter. To ensure this, the organic phase discharged from the decanter is measured, and as soon as a set limit is underrun, the airlift valve of the mixed phase is opened a bit more.

5.2 HS1 COLUMN

In the HS1 column, the organic HAP volumetric flow, which contains a fraction of extracted fission products together with the uranium, is stripped of these fission products by washing with diluted nitric acid (Fig. 6). Like the HA column, also this column is operated in the aqueous disperse mode.

At the top of the column, the HS1S scrub flow is fed together with the recycled HS2R flow from the HS2 mixer-settler. The organic HAP volumetric flow is fed to the bottom of the column.

The phase ratio is set by the ratio of HAP/HS1S + HS2R.

The reference variable for this ratio control is the HAP volumetric flow coming form the HA column.

Depending on the magnitude of the HS2R volumetric flow rate the HS1S volumetric flow rate is then matched so that the phase ratio remains constant.

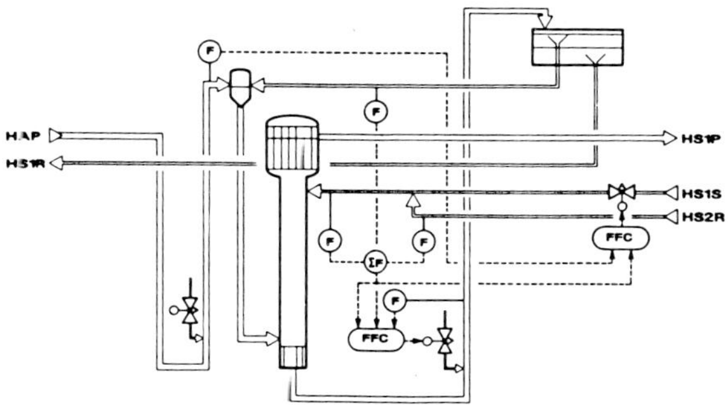

Fig. 6: The HS1 column of PLURA

Mixed-phase control works like the control system explained above for the HA column.

WAK experience in work safety and radiation protection

W. Kraut

SUMMARY

An overview is given covering essential aspects of work being done by the WAK work safety and health physics service.

WAK experiences, exspecially in routine health physics work, dosimetry and stack monitoring are presented and illustrated by diagrams with yearly figures.

Dr. W. Kraut, Wiederaufarbeitungsanlage Karlsruhe, Betriebsgesellschaft mbH, Postfach 2 20, 7514 Eggenstein-Leopoldshafen 2, Federal Republic of Germany

1 ACTIVITIES IN THE WAK WORK SAFETY AND HEALTH PHYSICS SERVICE

During the more than 15 years since reprocessing started at WAK, work safety and health physics aspects have changed considerably.

In this context the new regulations concerning radiation protection, immission protection and accident prevention and also the large number of recommendations on a national (for example BMI, KTA, DIN etc.) and an international level (for example ICRP, ICRU, ISO etc.) come to mind.

To meet these requirements, the health physics service has to be strengthened permanently with respect to personnel, apparatus, equipment and organisation.

The WAK health physics service is completely integrated in the plant operating group.

The following typical activities can be specified:

- planning and surveyance of any work in which safety and radiological aspects must be considered
- dosimetry
- stack monitoring
- routine area monitoring
- radioactive transports
- routine inspection of monitoring devices
- supply of protective (clothing)
- task force.

2 WAK EXPERIENCE

2.1 RADIATION PROTECTION AND ADEQUATE PLANT LAYOUT

To reduce the potential risk of being contaminated and to guarantee effective control measures a clear decoupling of all traffic routes for workers, visitors and material transportation has to be established. In a pilot plant like WAK with direct maintenance working conditions and ca. 5000 visitors a year, this is not realizable in all plant areas by plant layout and therefore has to be maintained by administrative regulations.

Under real working conditions room ventilation has to be properly controlled. Advantage can be taken of the fact that in the WAK air is generally conducted from rooms with a lower contamination potential to rooms with a higher contamination potential by pressure gradients up to 700 Pascal.

Problems can often arise with low level measurements. Obviously such measurements cannot be carried out in all plant areas because of a too high background radiation. It is therefore an urgent need for the health physics service to have rooms available with very low radiation level, a condition not easily fullfilled for rooms in the controlled area.

2.2 PRACTICAL ASPECTS

Any work in radiation fields or in contaminated areas has to be carefully planned and coordinated among the groups involved such as maintenance, operation and health physics groups.

There is a priority for having preventive protection measures. If there is any lack of information on actual working conditions, this has to be balanced out by preventive measures.

To achieve high quality standards, protective clothing should be standardized and continuously improved. In the course of time there has evolved a certain classification of rooms according to which a standard protective clothing and equipment has to be worn.

In this standardised equipment for example transportable decontamination showers are no longer in use. If necessary, a dry fixation of radioactivity shows better results.

There is also a strong need of standardization and optimization with respect to measuring techniques and measuring devices. Only then can misinterpretation of data be avoided.

Under these aspects WAK can show a positive development in health physics and safety work. In figures 1 to 3 WAK specific data for occupational accidents and radiation exposures are given.

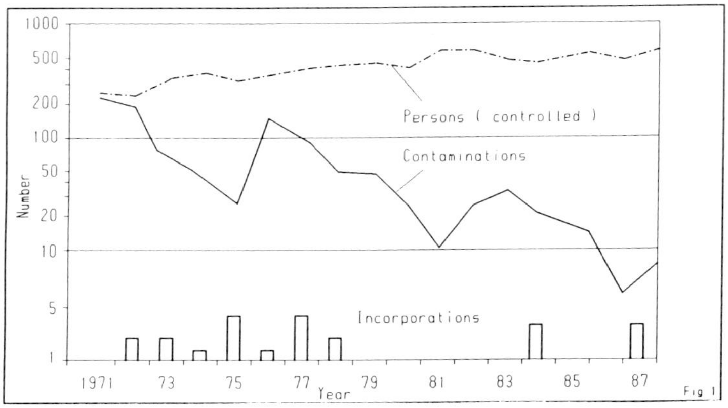

Fig. 1: Contaminations and incorporations in the WAK plant (Staff and contractors)

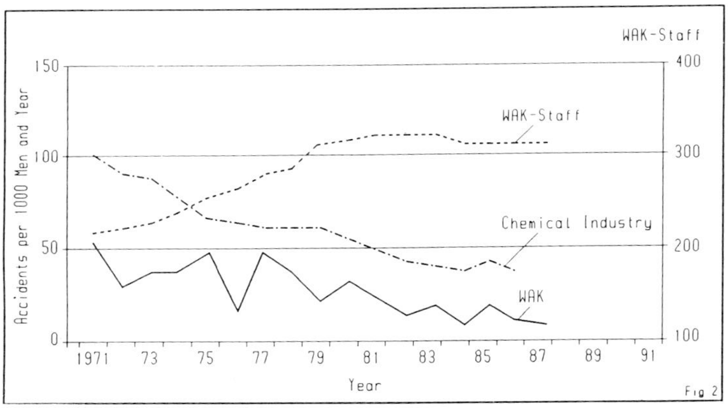

Fig. 2: Occupational accidents

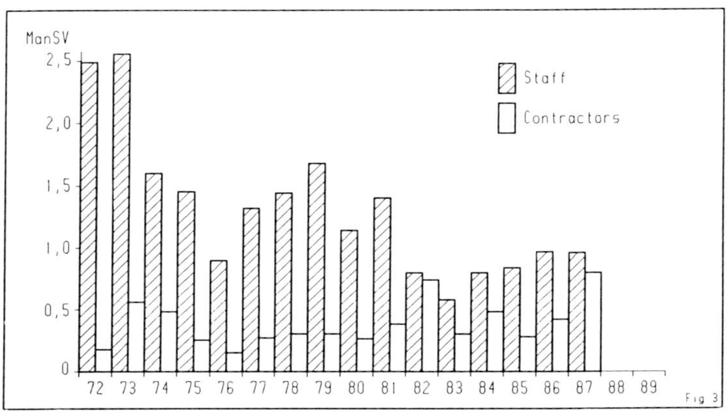

Fig. 3: Cumulative dose rates in the WAK plant

2.3 DOSIMETRY

In the WAK extensive dosimetry is carried out for external β-and γ - radiation.

The film dosimeter is the official personal dosimeter. A series of intercomparisons for various personal dosimeters conducted at the WAK in 1982 showed good agreement for penetrating γ-radiation between the film and a thermoluminescent dosimeter (TLD) encapsulated with 500 mg/cm^2. With 88 % of the individual dose values below 1 mSv a correlation coefficient between film and TLD of r = 0,95 was obtained.

In our opinion adequate β-dosimetry is only possible with TL-detectors.

Because of their high β-particle energies, β-emitters in reprocessing like Sr/Y-90 (E_{max} = 2,3 MeV), Ru/Rh-106 (E_{max} = 3,5 MeV) and Ce/Pr-144 (E_{max} = 3,0 MeV) have to be considered not only in skin dose evaluations but also in organ dose assessment (/1/).

Good results in determining β-field dose-rates were obtained with a large ionization chamber from Nardeux (Type Babyline 31/81). Yet under difficult working conditions the Babyline showed robust performance and accurate readings.

Specific n-dosimetry might be restricted to special working areas only: Pu-tail-end and fuel element reception hall. Actual n-dose rates are typical < 20 % of the prevailing -dose rate.

Individual n-dose can be estimated to range well under 1 % of the annual dose limits. In the WAK it is planned to put an Albedo-dosimeter system into operation in the near future. It is then hoped to exclude experimentally any significant n-dose percentage.

To have an efficient automatic and computer compatible dosimetry system WAK will replace the current personal pen dosimeters by the ALNOR electronic dosimetry system RAD 90.

2.4 STACK MONITORING

WAK is one of several emitters of the Karlsruhe Nuclear Research Center.

At the WAK site there are 3 stacks where the effluent activities have to be controlled.

The authorized annual discharge limits are shown in table 1, daily discharges are limited to 1 % of the annual values.

Continuous monitoring is carried out for gross α- and β-aerosol radioactivity, I-131 and the rare gases (Kr-85 being the only relevant nuclide). All other nuclides are controlled by a special sampling scheme and subsequent analysis in the laboratory.

As there is no retention for the two gaseous effluents Kr-85 and C-14 in the WAK off-gas cleaning system, the radioactivity emitted can be calculated in advance by the KORIGEN Code.

A comparison between calculated and measured values shows that there is a conservative figure with respect to the calculated emissions.

For C-14 measured values are on average 70 % of the calculated results, for Kr-85 there is a figure of about 90 %, respectively.

Another point worth mentioning is a special control scheme for the daily Pu-241 stack discharge. Since Pu-241 is not readily detectable because of its low β-energy of E_{max} = 20 keV, the Pu-241 discharge is estimated by taking

$$\text{activity Pu-241} = f \times \text{activity gross } \alpha$$

where f is a factor determined for each reprocessing campaign either by calculation according to KORIGEN or by mass spectroscopic analysis of the feed solutions as

$$f = \frac{\alpha \text{ activity Pu}}{\beta \text{ activity Pu}}$$

This procedure proved to be a conservative estimate. The Pu α-activity is clearly overestimated by the gross α-activity value.

In general it can be confirmed that there are precise methods to control the stack effluents accurately far below the daily discharge limits.

In Table 1 the actual 1986 stack releases are shown. The aerosol values decreased to very low values compared with previous years. Exploitation of time correlations between process operations and stack activity monitoring lead to an optimization of purge air ventilation and to a marked reduction of aerosol source terms - a promising result of the joint efforts by the WAK operation, research and health physics groups.

Nuclide	authorized annual limit [Bq]	actual discharge 1986 in % of annual limit
α-aerosol ($t_{1/2}$>8d)	$3,7 \times 10^{8}$	1,33
β-aerosol ($t_{1/2}$>8d)	$7,4 \times 10^{10}$	0,58
rare gas	$1,3 \times 10^{16}$	24,36
HTO	3,7 x 10	20,93
C-14	$6,1 \times 10^{11}$	30,05
Sr-90	$3,7 \times 10^{9}$	0,45
I-129	$2,4 \times 10^{8}$	75,54
I-131	$1,48 \times 10^{9}$	9,40 **)
I equiv.*)	$1,48 \times 10^{9}$	83,65 **)
Pu-241	$7,4 \times 10^{9}$	2,08

*) $\text{I equivalent} = \text{I-131} + \frac{\text{I-129}}{0,165}$

**) 6,4 % of the stack release of I-131 is a consequence of the Tschernobyl accident.

Table 1: WAK stack discharges

3 PERSPECTIVES

There is an ongoing development in radiation protection work with respect to technical, methodological and scientific questions, only to mention for example the implementation of microprocessor technology in measuring devices, biological dose assesment etc.

Training, education and information remains a permanent need for all workers in the field of health physics so as to provide a continuous degree of competence.

In Germany we are still awaiting a new issue of the radiation protection ordinance.

4 LITERATURE

/1/ Bundesanzeiger Nr. 126 vom 12. Juli 1988, pp. 21.28: "Berechnungsgrundlage für die Ermittlung von Körperdosen bei äußerer Strahlenexposition durch Elektronen, insbesondere durch ß-Strahlung" (Empfehlung der Strahlenschutzkommission)".

Part II
Waste Treatment and Storage

On the process chemistry of HLLW-vitrification in a liquid-fed ceramic melter (LFCM)

H. Pentinghaus

SUMMARY

For the first time a comprehensively simulated HLLW having the full content of the noble metals Pd and Ru has been vitrified by means of a LFCM on a technical scale yielding a glass product homogeneous with respect to the glass matrix and the dispersion of the insoluble noble metal compounds RuO_2 and Pd-tellurides. The vitrification of 5.5 m^3 of the waste simulate led to about 2.5 tons of glass product. However, considerable agglomerations of RuO_2 and liquid Pd-tellurides occurred in the bottom region of the LFCM-system which could be remobilized by bubbling an inert gas through the melt during its drain. Minor amounts of complex salts, being relics of the vitrification process, could be traced in some parts of the glass product. Analyses performed reveal the basic processes involved which are leading to the agglomerations.

Prof. Dr. H. Pentinghaus, Institut für Nukleare Entsorgungstechnik (INE), Kernforschungszentrum Karlsruhe GmbH, Postfach 36 40, D-7500 Karlsruhe 1, Federal Republic of Germany

1 INTRODUCTION

Principally the vitrification of high level liquid waste (HLLW) consists of a dissolution of the waste oxides in a borosilicate glass melt. Pouring the melt into canisters and their subsequent cooling finally results in the solid waste form.

The noble metals are a class of elements which are of special importance in the HLLW, because they are not substantially soluble in any glass melt. Furthermore they form complex compounds involving oxygen, - $(Ru,Rh)O_2$ - and Te - (Pd,Rh)-tellurides - under the conditions of vitrification. Both compounds show metallic conductivity and the tellurides are at least partly liquid at process temperatures. MoO3, another substantial waste constituent has quite a low solubility in the glass melt so that it determines the waste loading of the final waste product. In addition, the dissolution kinetics of MoO_3 may temporarily give rise to the formation of salt melts.

Thus, developing a technology for the vitrification of HLLW arising from the reprocessing of spent light water reactor (LWR) fuel elements inevitably means to get insight into the chemical details of the process by performing tests on technical scale first under non-radioactive conditions by using waste simulates. This indeed is a point of major concern, because application of inadequate simulates may give rise to misinterpretations and may avoid any reliable predictions with respect to the behaviour of HLLW during the vitrification process. As an example, for the LFCM-system now available, larger amounts of conducting metallic particles in the glass melt may heavily interfere with the high strength electrical fields causing a decrease of homogeneous dissipation of heat. For the study of this effect a first vitrification campaign on technical scale has been performed to process a comprehensively simulated HLLW with full content of the noble metals Ru and Pd, resulting in a glass product having a formal waste loading of 15 wt. %.

2 PREREQUISITES OF THE VITRIFICATION

The basic prerequisites of the vitrification of HLLW comprise the melter technology, the waste simulate and the basic glass. The technique applied has been developed at INE /1/ and consists of a one-stage liquid-fed ceramic melter (LFCM) system which is continuously fed with HLLW and discontinuously with glass beads. This means that evaporation, calcination and vitrification occur simultaneously. The mean residence time of the product melt is about 24 h which is more than sufficient for complete homogenization as evidenced from experimental results. The glass melt is released semi-continuously through a bottom drain. The main characteristic of the LFCM is its specific temperature distribution inside the system with a "cold cap" covering about 90 % of the melt's surface. It is just here where most of the chemical reactions as well as physical operations take place which comprise the vitrification process. Fig.1 schematically shows the important features of the LFCM.

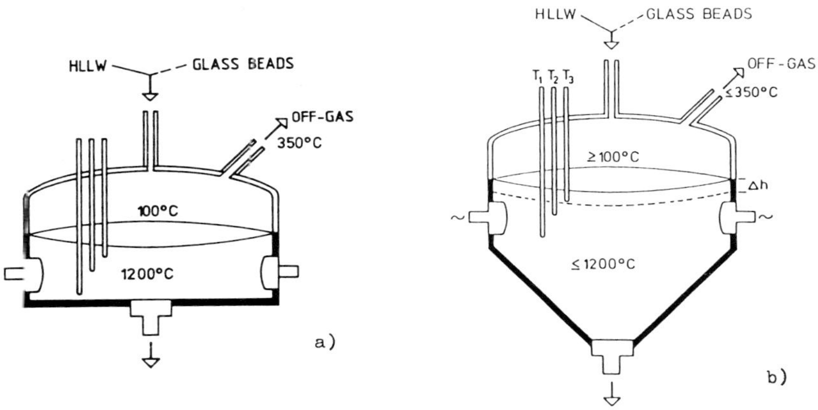

Fig. 1: Scheme of the one-stage LFCM: a) flat bottom, b) sloped bottom configuration

The HLLW simulate is based on the reference data given for the reprocessing plant WAA-Wackersdorf, now under construction. These data are presented in Fig. 2 showing the concentrations in g/l. The specific significance of this HLLW is its very low concentration of corrosion products. The simulate contains all the elements shown in Fig. 2 except for the actinides, Tc and Rh. Only the latter has been substituted by Pd for economical considerations. So the formal waste loading of 15 wt. %, based on the reference data, is actually only about 12 wt %.

The basic glass, coded VG 98/12.2, for the final waste form is based on borosilicate glasses the development of which has been conducted empirically as in other countries running a HLLW management program. VG 98/12.2 is produced by an optimized recipe derived from the parent glass VG 98/12 /2/ which has been thoroughly characterized some years ago by INE. A compositional variation became necessary due to new requirements to meet a lower viscosity function which arose from the experiences gained from the operation of several melter systems vitrifying chemically different HLLW simulates.

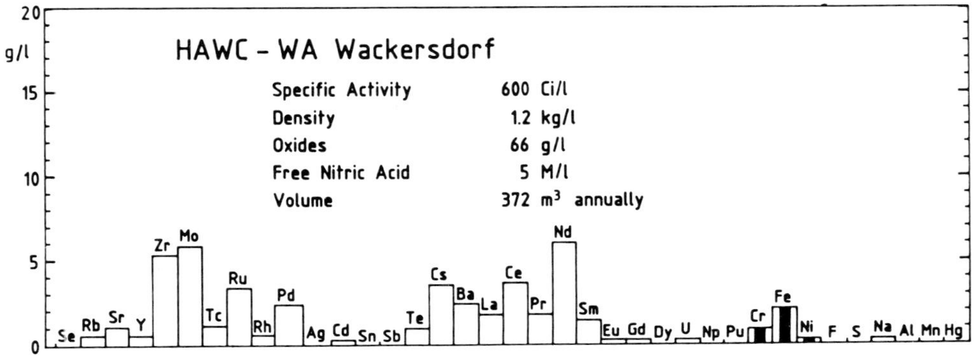

Fig. 2: The composition of the HLLW arising from the future reprocessing plant at Wackersdorf

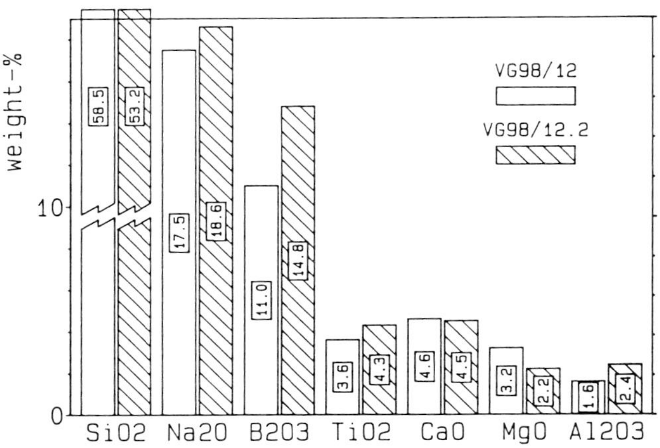

Fig. 3: Compositional differences between the basic glasses VG 98/12 and VG 98/12.2

The changes in composition did not affect electrical conductivity and hydrolytic stability significantly, these properties maintained in the same order of magnitude. In Fig. 3 the compositional differences between the basic glasses VG 98/12 and VG 98/12.2 are shown.

3 QUALITY OF THE WASTE FORM

The continuous vitrification process applied during the campaign under discussion will not include under industrial conditions any sampling and analyzing of the final waste form because its quality is guaranteed by process control. In detail the product quality is governed by the following factors:

- Recipe of the basic glass;

- Process control, which comprises the form of the basic glass and its grain spectrum, feed rates of HLLW and/or of basic glass, extent of the "cold cap" covering the glass melts' surface, oxygen fugacity in the melter system;

- Thermal treatment of filled canisters to minimize internal surface increase by cracking.

The features of product quality as such can be summarized under four items:

- Homogeneity of the glass matrix, depending on process parameters like form and size of the basic glass (beads) and residence time. The homogeneity is reduced by inherent phase transformations, provided their kinetics are sufficiently fast, as liquid/liquid phase separation and/or devitrification. Both are dependent on the recipe of the basic glass as well as on the thermal history of the waste form;

- Dispersion of insoluble constituents RuO_2 and Pd (Rh)-tellurides also due to process factors, residence time, ad form of the basic glass;

- Increase in surface area through cracking in the filled canisters depending on thermal history of the canisters;

- Hydrolytic reactivity and kind of alteration products which are functions of the basic glass recipe and the repository chosen.

4 DETAILS OF THE CAMPAIGN

The vitrification of 5.5 m^3 of HLLW simulate by means of the LFCM system VA-2 at INE has been performed with a mean feed rate of 15 l/h for the HLLW simulate. Glass beads were fed having a grain spectrum between 0.1 mm up to 1.25 mm in diameter. About 2.5 tons of glass product have been produced homogeneously with respect to the matrix and the dispersion of insoluble RuO_2 and Pd-tellurides. Whereas RuO_2 crystals are always smaller than 1 um the tellurides turned out to be slightly larger with a mean diameter of 5 um and their composition varying between 92Pd.8Te and 80Pd.20Te.

In some parts of the glass product minor amounts not exceeding about 200 ml of a complex salt, the composition of which is shown in fig. 4 could be traced. It was of yellow colour due to the presence of hexavalent chromium and apparently a relic of the vitrification process. Of much greater importance was the fact that 67 wt.% of Ru and 53 wt.% of the Pd remained in the melter system as a dilute suspension of RuO_2, Pd-tellurides and glass melt having an increased viscosity. Due to the flat melter bottom and a slightly elevated drain hole the suspension could not flow out at the end of the campaign. The accumulation of insoluble compounds obviously influenced the electrical resistivity in the lower electrodes' area which continuously decreased (at constant current). These interactions, however, did not result in a breakdown of the electrical system.

5 ANALYSES

A sequence of individual operations constitutes the process of vitrification. Though all steps occur simultaneously in the "cold cap" for every individual volume element the order of sequence is as follows:

- evaporation of H_2O/HNO_3,
- desiccation, and with increasing temperature,
- eutectic melting of the nitrates,
- dissolution of molybdenum oxide,
- thermal decomposition of nitrates and nitrato complexes and finally, above the softening point Ts (about 600° C) of the basic glass,

- temporary survival of alkali molybdate melts.

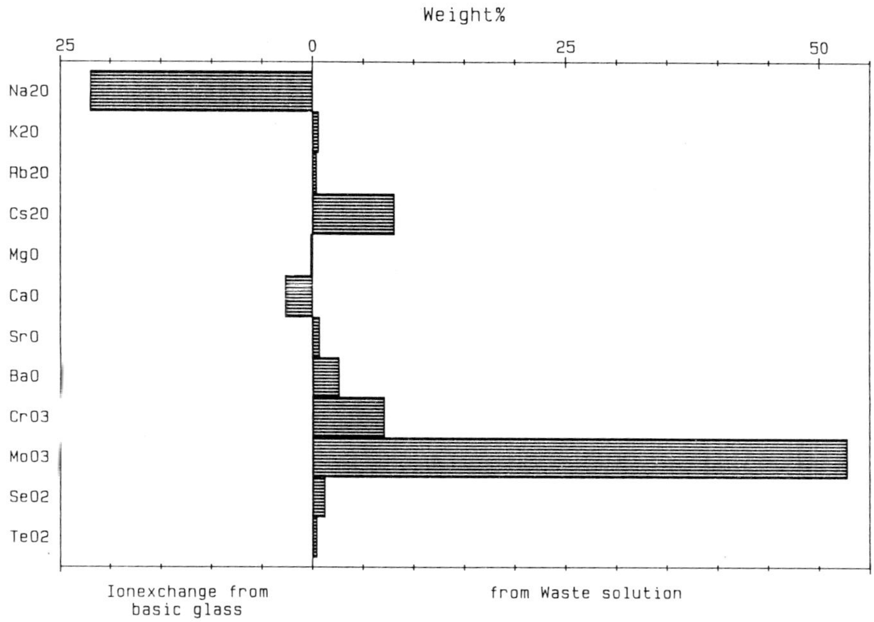

Fig. 4: Typical composition of the "yellow phase" from the HLLW vitrification process. Sources of constituent oxides are indicated

These reactions happen in close contact with the basic glass beads which do not stick tightly together despite the fact that they are densely packed. Degassing is easily achieved. Above T_S the interstices between the beads are closed and the remaining decomposition products as well as the molybdate melts are included in the glass melt. These tiny inclusions will be transported into the hot areas of the melt by convection and thereby dissolved completely. It has been found that application of a certain size of glass beads is most effective for a continuous vitrification avoiding physically any large agglomeration of molybdate based melts as well as of noble metal compounds.

The complex ionic liquids which exist temporarily during the vitrification process in the "cold cap" region of the melter system for kinetic reasons result in the frequently quoted "yellow phase" in the solid state. Its nature has been clearly defined by now. Regarding its composition, it is based on alkali molybdates and its formation is due to the thermal stability of the parent Na_2MoO_4-melt which decomposes at about 1400°C only. This is far above the highest temperature in the LFCM as shown in Fig. 5 together with data of $NaNO_3$-, $NaNO_2$- and chromium molybdate melts. The kinetics of the dissolution of such an ionic melt are mainly determined by the limited solubility of Mo the saturation concentration of which controls the waste loading and is about 2.1 wt.% at a target value of 1.95 wt.% Mo in the glass product.

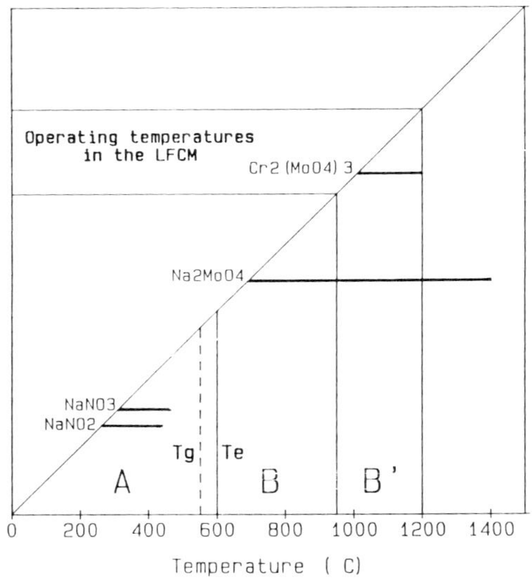

Fig. 5: Thermal stability of some basic ionic melts formed during the vitrification process

The composition of the ionic liquids changes with their life times. In the beginning they can only be formed from the alkali cations and MoO_3 present in the HLLW. Complex ion exchange reactions with the glass beads starting as soon as nitrate melts are present lead to an enrichment of Na, Ca and Mg. A typical composition of its final state is given in Fig. 5 where a distinction is made between the sources of the constituent oxides via ion exchange from the basic glass and from the waste solution. As a certain amount of the chromium present is oxidized to the hexavalent state it makes the complex salts yellow. Larger agglomerations of such an ionic liquid in the LFCM having a higher density than the glass melt in contact with it will sink to the bottom of the melter against the surface tension which at higher dispersion may keep it on top.

The strong temperature dependence of its dissolution in the glass melt may then become responsible for its survival. However, a higher dispersion of these ionic liquids by feeding appropiate glass beads safely avoids agglomerations of this kind.

The noble metals Ru, Pd and Rh present in the HLLW will not be dissolved substantially in the glass melt. As expected RuO_2 is formed; Pd and Rh, however, form tellurides in contrast to expectation. Both compounds are metallic conductors, and the tellurides are liquid down to about 800° C, as we know from a study of the system Pd-Te in its Pd-rich part which is currently being performed. This explains the spherical morphology and flow structures of the Pd-tellurides observed in the agglomeration at the bottom of the LFCM. In Fig. 6 first results of the study of the system Pd-Te are given, the arrows mark the compositional variation observed in the agglomeration. The triangle indicates the highest temperature in the LFCM. The homogenously distributed tellurides determined in the glass product have a lower Te-content which is close to the solid solubility of Te in Pd at about 800° C and it is assumed that they have been formed primarily.

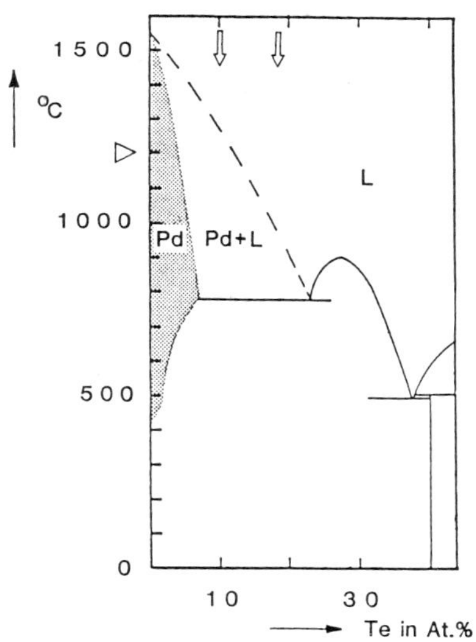

Fig. 6: Schematical phase diagram of the Pd-rich part in the system Pd-Te

There is strong evidence that elemental interactions and changes in oxigen fugacity (fO_2) are responsible for the observed compositional differences.

When such an agglomeration of noble metal compounds has been formed in a LFCM with a flat bottom - as in this campaign - the question arises how to remobilize it. This has successfully been achieved by bubbling inert gases through the melt during its draining. The chemical composition of the tellurides did not change during this operation.

In order to remove the dilute noble metal suspensions regularily by bottom drain future melter systems will have a sloped bottom configuration as shown in Fig. 1b.

6 OUTLOOK

The basic principles of the vitrification process are known today /3/; some details, however, are still awaiting explanation. As many elements in the HLLW can exist in different oxidation states, their mutual interactions as well as their redox pair yields as a function of fC_2 which are unknown so far determine the vitrification process substantially. Their study will be the main task for the future. Measurements of individual redox pair yields as a function of oxygen fugacity are of high priority and will provide the insight we need. The behaviour of Tc may be taken as an example. In the 7-valent state Tc_2O_7 will be formed and be driven into the off-gas system. In the 4-valent state, however, it will be dissolved in the melt and become part cf the glass structure. Probably the vitrification of HLLW can be achieved with no extra redox control.

LITERATURE

/1/ Weisenburger, S.,
IEEE Transactions on Industry Applications,
IA-18, p. 73-82 (1982)

/2/ Guber, W.; Hussain, M.; Kahl, L.; Ondracek, G.;
Saidl, J.; Dippel, Th.,
Report KfK 2721 (1981)

/3/ Pentinghaus, H.;
Report KfK 4177 (1987), p. 148 - 185

Long-term studies on the leachability of cemented and non-cemented iodine-129 loaden sorption material

R. Kaempffer, J. Furrer

SUMMARY

Leaching tests on the loaded AC 6120 iodine sorption material (12 wt. % Ag) in water and salt brines were performed over a rather long period of time to allow better judgement of the behavior of radioactive waste disposed of in a salt dome.

The utilization of capacity of the loaded iodine sorption material from the Karlsruhe reprocessing plant (WAK) was 95 % related to the amount of silver added.

The result of the stationary leaching tests has been a leaching rate of the material not embedded in cement of < 0.1 %, whereas the leaching rate of the iodine sorption material embedded in cement has been < 0.01 % of the total iodine-129 inventory.

After addition of carbon steel to the sorption material embedded in cement the same leaching rates were measured as for material not embedded in cement. The addition of stainless steel exerted but little influence on the leaching rate.

R. Kaempffer, Dr. J. Furrer, Kernforschungszentrum Karlsruhe GmbH, Laboratorium für Aerosolphysik und Filtertechnik II, Postfach 36 40, D-7500 Karlsruhe 1, Federal Republic of Germany

An essential step in conditioning radioactive waste prior to disposal consists in solidifying and subsequently embedding it in a stable matrix so that water or brines which might attack it during disposal will not cause leaching of a major portion of the radioactive inventory.

The fission product iodine released during dissolution of spent fuel elements in the Karlsruhe reprocessing plant is removed from the dissolver offgas by a gas-solid reaction involving iodine and silver nitrate by which heavy soluble silver iodide is formed /1/. The iodine sorption material is prepared by impregnation with silver nitrate on a carrier material consisting mainly of high-annealed amorphous silicic acid (AC 6120/12 wt.% Ag).

As this material is envisaged for use in fission product iodine removal in the future Wackersdorf reprocessing plant (WAW) the exhausted iodine filter elements have to be cemented in 400 l drums and stored in a salt stock or in a mine.

It was the goal of the investigation to determine the iodine leachability on cemented an non-cemented, iodine-129 loaden sorption material using test parameters which resembled practical conditions. Another objective was to study the influence on leachability of the iron hooped drum and canister packaging materials (ordinary steel and stainless steel, respectively) (electrochemical series).

The following solutions were used under the leaching program for the AC 6120 iodine sorption material:

a) saturated NaCl solution (about 27 %)
b) carnallite solution according to the following formula:

KCl	:	3.125 g	=	0.6 %
NaCl	:	1.060 g	=	0.2 %
$MgSO_4 \cdot 7H_2O$	:	20.900 g	=	3.8 %
$MgCl_2 \cdot 6H_2O$	:	366.700 g	=	66.6 %
H_2O	:	158.215 g	=	28.8 %

$\delta = 1.325\ g \cdot cm^{-3}$ at 20° C.

8 g of loaden iodine sorption material with an I-129 total inventory of 1.76×10^6 Bq were prepared, cemented or non-cemented, with 20 ml leaching solution. Defined specimens of stainless steel and of ordinary steel (1.4571 and St 1203, respectively) were added to the cemented and the non-cemented preparations. The amounts of ordinary steel and stainless steel specimens, respectively, added were equivalent to the ratio of drum inner surface to material content.

I-129 was determined with an Intrinsic Germanium Detector supplied by Princeton Gamma Tech (PGT) (Resolution: 325 eV at 5.9 keV and 450 eV at 40 keV).

The determination makes use of the X-radiation emitted by the I-129 daughter nuclide Xe-129 (29.5 and 29.8 keV).

The I-129 detection limit of the measuring configuration is 1 ± 1 Bq per ml (1σ-error).

SUMMARY OF DATA FOR THE DETERMINATION OF THE STEADY-STATE LEACHING RATE OF IODINE-129

Total inventory:

8 g AC 6120 (12 wt.% Ag) $\hat{=}$ 1760.6 KBq iodine-129 (WAK material)
1 Bq iodine-129 $\hat{=}$ 1.53×10^{-7} g iodine-129
1760.6 KBq iodine-129 $\hat{=}$ 1.044×10^{-3} mole iodine-129

Iodine-129 concentration

mmole iodine-129/g dry cement: $\frac{1.044}{15} = 0.07$

Type of cement:	Portland cement
Water/cement/weight ratio:	9 g/15 g = 0.6

Data on the cement monolith:

Height	3.3	cm
Diameter	2.6	cm
Volume (V)	17.52	cm^3
Surface (S)	27.0	cm^2
V/S	0.65	cm

The dimensions of the leaching vessel were chosen in such a way that the ratio

$$\frac{\text{volume of leaching agent}}{\text{surface of specimen}} = \frac{20\ cm^3}{27\ cm^2} = 0.74\ cm$$

The results of the investigation are visible from Fig. 1 where the steady-state iodine leaching rate of the sorption material has been plotted as a function of the duration of leaching.

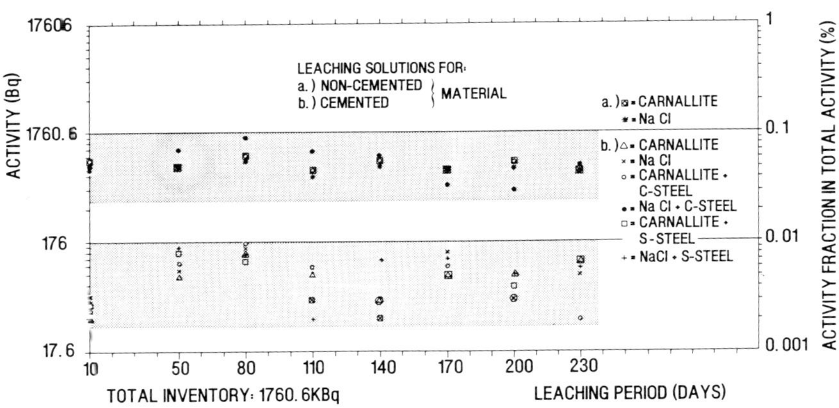

Fig. 1: Iodine-129 leaching rate for cemented and non-cemented iodine sorption material AC 6120 as a function of the duration of leaching

Two characteristic areas can be distinguished:

a) Leaching rate of the cemented iodine sorption material $< 10^{-2}$ % of the I-129 total inventory.

b) Leaching rate of the non-cemented iodine sorption material $< 10^{-1}$ % of the I-129 total inventory.

ad a) Adding stainless steel does not exert any influence on the leaching rate.

ad b) Adding NaCl solution + ordinary steel produces a behavior similar to that of non-cemented material.
Reason: potentials of the electrochemical series

$$Fe \rightleftharpoons Fe^{2+} + 2e = -0.44\ V$$
$$Ag^{+} + e \rightleftharpoons Ag = +0.80\ V$$
$$\Delta = 1.24$$

Cementing of the iodine sorption material and emplacement in a borehole are recommended for disposal. In addition, it is proposed to use a stainless steel waste canister accommodating the iodine package for disposal in 400 l waste drums.

LITERATURE

/1/ WILHELM, J.G.; FURRER, J.; SCHULTES, E.:
"Head-End Iodine Removal from a Reprocessing Plant with a Solid Sorbent."
CONF 760 822 (1977) Vol. 1, p. 447

/2/ FURRER, J.; KAEMPFFER, R.; WILHELM, J.:
"Management Modes for Iodine-129 Radioactive Waste and Management."
ISSN 0275-7273 (1982), Vol. 7, pp. 275-288

Determination of sorption data of radionuclides with site-specific samples

G. Tittel, A. Hollmann, E. Warnecke

SUMMARY

For the planned German repositories, i. e. the former iron ore mine Konrad and the mine in the Gorleben salt dome, the long term safety has to be proved by a site-specific safety analysis. As it is anticipated that releases of radionuclides in the postoperational phase can only occur with water as a transport medium, it is necessary to evaluate the radionuclide retardation by the geosystem. Therefore experimental investigations of the sorption behavior of radiologically relevant nuclides in natural water-sediment systems of the two sites using batch, column and diffusion techniques have been carried out under the auspices of the Physikalisch-Technische Bundesanstalt. The experimental conditions in the laboratory have been adjusted as closely as possible to the natural geological and hydrogeological situation. The dependence of the sorption data on parameters which have been identified to influence the sorption behavior has been extensively studied. The measured sorption and desorption coefficients especially show the strong influence of natural (e. g. humic substances) and technical complexing agents (e. g. EDTA).

Dr. G. Tittel, A. Hollmann, Dr. E. Warnecke, Physikalisch-Technische Bundesanstalt Braunschweig, Postfach 33 45, 3300 Braunschweig, Federal Republic of Germany

1 INTRODUCTION

In accordance with the Federal Republic of Germany's concept for the disposal of radioactive wastes in deep geological formations, the following repository projects are at present being pursued:

1. The former Konrad iron ore mine is being considered for the disposal of radioactive wastes with a negligible thermal impact upon the host rock. The above-ground and underground site investigation programme has been completed for this site. The Physikalisch-Technische Bundesanstalt (PTB) completed the plan approval documentation and handed the documents over to the licensing authority in 1986. The documents are at the moment under scrutiny by experts appointed.

2. The Gorleben salt dome is being considered for the disposal of all kinds of solid and solidified wastes. The shaft sinking is under way for this repository site.

The PTB has also started planning a facility for the deep storage of waters containing tritium.

As the applicant in the nuclear licensing procedure, the PTB must demonstrate that possible radiation exposures remain below the legal limits in the operational and in the post-operational phase of a repository.

In accordance with the safety criteria for the disposal of radioacitve wastes in a mine recommended by the German Reactor Safety Commission and published by the German Federal Minister of the Interior /1/,

- the necessary safety of a repository mine in a geological formation must be demonstrated by a site-specific safety analysis which takes into account geological relationships, repository mine and waste products / packages in the total system;

- during radionuclide releases from the repository mine, overlying strata and host rock must contribute to the hindering of inadmissible concentrations getting into the biosphere. Hence a high sorption capacity for radionuclides is favourable for the barrier function of the overlying strata and host rock to be fulfilled;

- even after decommissioning, radionuclides which might escape from the closed repository into the biosphere as a result of not entirely excludable transport processes may not lead to individual doses which exceed the values of section 45 of the German Radiation Protection Ordinance (30 mrem concept).

For both repository sites it is assumed that in the post-operational phase, the waste packages come into contact with saline formation waters which leads to a release of radionuclides out of the repository area into the surrounding rock formations.

The sorption properties of sediment / groundwater samples representative of the natural systems along the anticipated waterpath for radiologically relevant radionuclides have therefore been studied under the auspices of the PTB. These investigations have enabled model calculations to be made concerning radionuclide migration through the geosphere on the basis of experimentally estimated sorption data /2/. The laboratories which have been involved in the investigations are shown in table 1.

2 ORIGIN OF SAMPLES AND EXPERIMENTAL PROCEDURES

Up to now the sorption experiments with Gorleben samples corresponding to the envisaged migration scenario have been carried out with material from the above-ground exploration. As can be seen from table 2, these are chiefly sandy and consolidated sediments from the Quaternary and the Tertiary. The sorption / desorption experiments are being continued with samples from the shaft sinking, which is part of the subsurface exploration programme.

In experiments with consolidated material where a sufficient amount of formation water was not available, model waters derived from the analysis of the pore water of the sediments were used.

Table 3 comprises the sediment / water systems which have been investigated within the scope of the sorption experiments for the planned Konrad repository. Besides quaternary samples the Upper Cretaceous, Lower Cretaceous - a formation with a very low permeability - and the Malm geological formations were thoroughly investigated corresponding to the possible migration paths.

INSTITUTE	ELEMENTS UNDER INVESTIGATION
TECHNISCHE UNIVERSITÄT MÜNCHEN, (TUM) INSTITUTFÜRRADIOCHEMIE D-8046 GARCHING	Pu, Am,Cm, Np, Tc
GESELLSCHAFT FÜR STRAHLEN- und UMWELTFORSCHUNG MBH (GSF) INSTITUT FÜR RADIOHYDROMETRIE D-8042 NEUHERBERG	Co, Sr, Pd, I, Cs, C, Ni, Mo, Ra
STAATLICHES FORSCHUNGSINSTITUT FÜR GEOCHEMIE (SFG) D-8600 BAMBERG	Pb, Ra, Th, Pa, U, Ac, Ni, Sr
TECHNISCHE HOCHSCHULE DARMSTADT (THD) FACHBEREICH ANORGANISCHE CHEMIE UND KERNCHEMIE D-6100 DARMSTADT	Co, Sr, Tc, I, Cs, Ce, Pr, Np, Mo
FREIE UNIVERSITÄT BERLIN (FUB) INSTITUT FÜR ANGEWANDTE GEOLOGIE D-1000 BERLIN	Se, Zr, Nb, Tc, Sn, Gd
BUNDESANSTALT FÜR MATERIAL-PRÜFUNG (BAM) D-1000 BERLIN	U, Pu, Am, Cm

Table 1: Research groups and elements under investigation

Here, intensive experiments have been performed with samples from the Oxfordium, the iron ore-bearing formation in which the repository will be constructed. Samples from the Cornbrash, a stratigraphic unit below the repository area, have also been investigated, as the Cornbrash has been identified as a potential aquifer.

The rock samples were taken from the area of the former iron ore mine and from quarries. Care was taken to obtain samples from facies the same as those in the mine and which were largely free of weathering.

A deep drilling also made it possible to take fresh samples of all relevant strata for the experiments. The corresponding formation waters were taken from wells, the shaft area and in the mine. In the

Stratigraphic unit	Lithology	Kind of groundwater	Depth of sampling (m)	No. of samples
Quaternary-Pleistocene	Medium to coarse sand	Fresh	3, 10, 30	3
	Fine to coarse sand	Fresh	20, 90, 130	3
	Fine to coarse sand	Saline	60	1
	Clay	Model water	50	2
	Boulder clay	Model water	40	1
	Silt	Model water	90	1
	Fine to medium sand	Fresh	110, 120	2
	Fine to medium sand	Saline	220	2
	Fine to coarse sand	Saline	200, 210, 240	4
	Silt to sand	Model water	155	1
	Silt	Model water	145	1
	Clay	Model water	130	1
	Medium sand	Saline	180	1
Tertiary	Fine to medium sand	Fresh	55, 100, 140	3
	Fine sand	Saline	75	1
	Silt	Model water	150	1
	Clay	Model water	200	1
	Fine sand	Model water	175	1
	Silt	Model water	170	1
	Clay	Model water	190	1

TABLE 2: Origin of samples from the Gorleben site for radionuclide migration experiments

deep drilling no water effluent allowing a sample to be taken was found. A model water based on pore water analysis had to be used for

Stratigraphic unit	Lithology	Kind of groundwater	Depth of sampling (m)	No. of samples
Quaternary	Sand	Fresh	60	1
	Sand and clay	Fresh	60	1
Cretaceous	Limestone	Fresh	Surface 110	2
	Limestone	Saline Model water	Surface 200	2 1
	Claystone	Saline	310, 420	2
	Sandstone	Saline	460	1
	Clay-marlstone	Saline	570	1
Jurassic	Limestone	Saline	Mine	1
	Iron ore	Saline	Mine	3
	Limestone	Saline	Mine	2
	Clay-marlstone	Saline	Mine	1
	Calcareous sandstone	Saline	890	2
	Claystone	Saline	940	1

Table 3: Origin of samples from the Konrad site for radionuclide migration experiments

the Cenomanian. Before describing the investigations and their results in more detail the correlation between sorption / desorption experiments and the model calculations of radionuclide migration will be illustrated:

- In the model calculations the barrier function of the stratigraphic units to be taken into consideration is characterized by element specific retardation factors which are a measure of the delay of a radionuclide against water as a transport medium as a result of sorption processes.

- The retardation factor connects density and porosity, the characteristic rock parameters, with the distribution coefficent

K_D of a radionuclide for the sediment-water-system under consideration. The K_D value is derived from the experimentally estimated sorption / desorption coefficents (R_S / R_D values) or can be directly obtained with dynamic experiments.

The sorption / desorption coefficients are obtained by measuring the distribution of radionuclides between the sediments and the corresponding waters of the host rock and overlying strata from the sites of the two planned repositories.

For this the following experimental techniques were used:

- Batch methods
  These are subdivided into:

  · Shaking experiments with slow and careful shaking (figure 1).

    Here, the influence of relevant parameters on the sorption behavior of radionuclides can be measured in relatively short times. Nevertheless, in some cases experiments taking 1 year or more proved to be necessary. Shaking experiments result in R_S / R_D values.

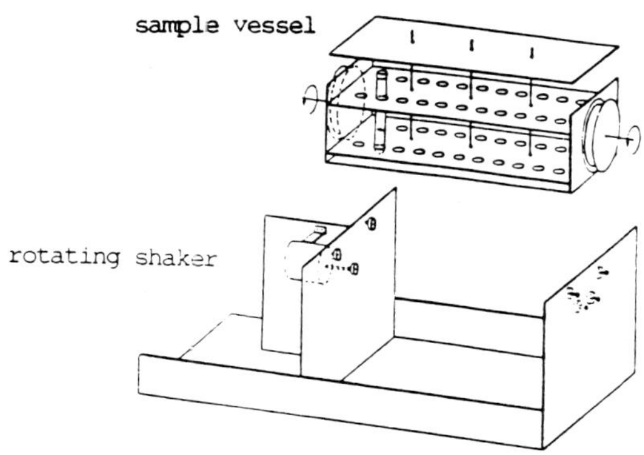

Fig. 1: Schematic representation of the shaking arrangement

  · Recirculation column experiments (figure 2).
    Here, the waters containing the radionuclides are circulated until constant distribution of radionuclides between the solid and liquid phase. Here too, R_S / R_D values are obtained.

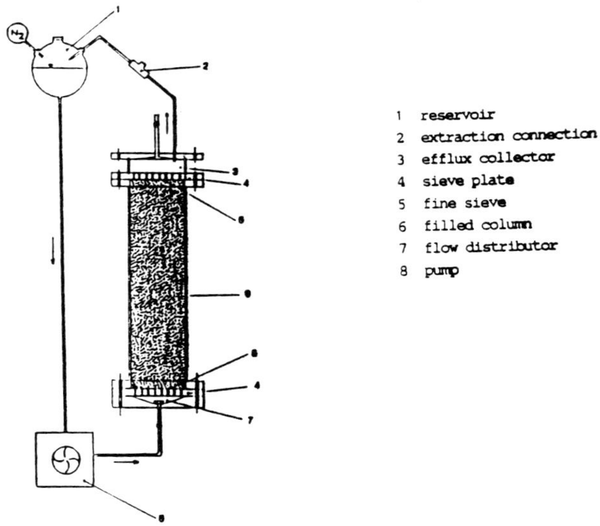

Fig. 2: Schematic representation of a recirculation column

- Flow-through column experiments for estimating dynamic transport processes (figure 3).

  Here, a solution either spiked or pre-equilibrated with radionuclides passes through a column filled with the respective sediment. K_D values for the radionuclides are deduced from the measured break-through curves.

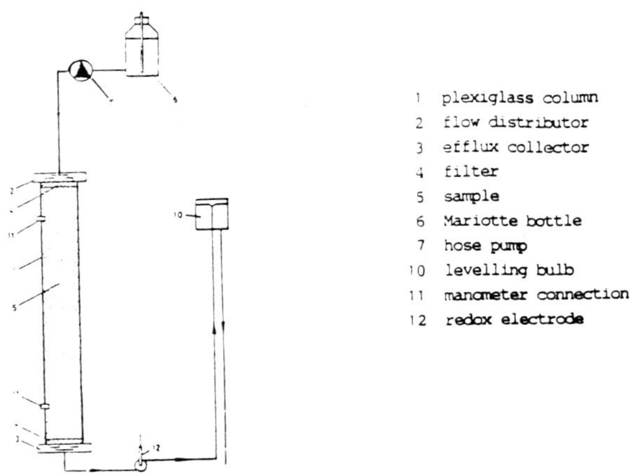

Fig. 3: Schematic representation of a flow-through column

- Diffusion experiments (figure 4)

  Diffusion experiments are carried out with consolidated material which as a rule has a low hydraulic conductivity. Here, diffusion can be regarded as the main driving force for the radionuclide transport. In these experiments the K_D value is derived from the measured data by means of transport equations.

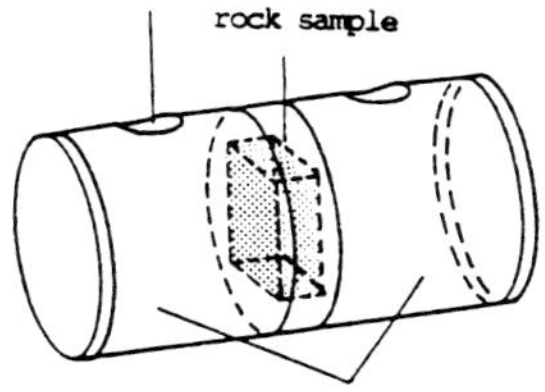

Fig. 4: Schematic representation of the arrangement used in diffusion experiments

As already mentioned, the results obtained with these fundamentally different techniques represent an important basis for the assessment of the long term safety for both the planned repositories. Experimental conditions in the laboratory should therefore approach the natural geological and hydrogeological situation at the respective sites as closely as possible. As the natural waters and sediments represent a very complex multi-component system, experiments with such samples require a thorough pretreatment of samples and long experimental times.

Figure 5 illustrates the procedure for conditioning Konrad samples for sorption / desorption experiments with actinide elements.

For the other elements and systems under investigation, similar conditioning procedures were applied.

Extensive experiments with a variation of parameters which have been recognized as relevant for the sorption behaviour of radionuclides have been performed. Table 4 gives a survey of the parameters investigated. On this basis it is possible to interpret the sometimes

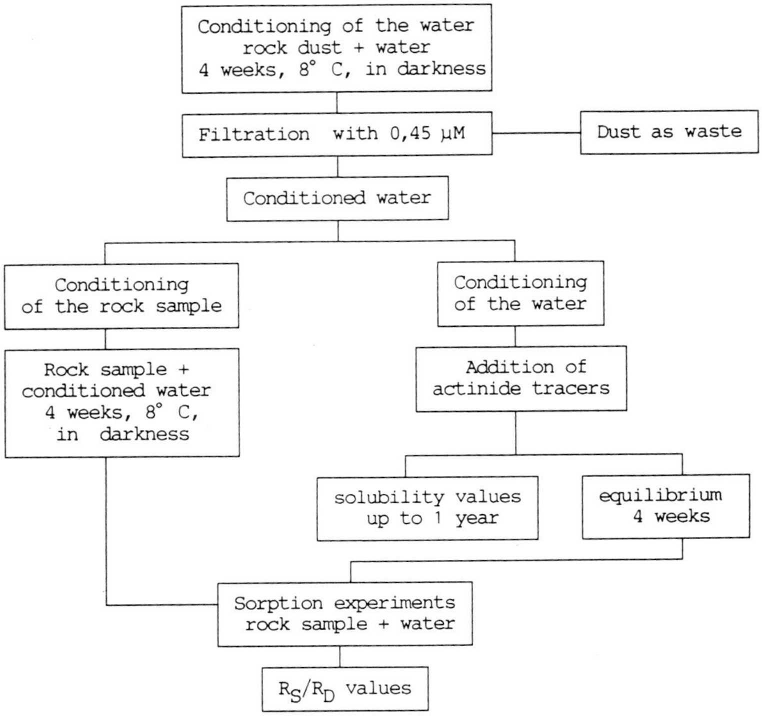

Fig. 5: Conditioning of the samples

widely varying sorption data of a radionuclide and to transfer laboratory data to the natural system.

3 RESULTS AND DISCUSSION

The investigations have revealed that:

- observance of the natural redox conditions in the laboratory is of particular importance for redox sensitive radionuclides like Tc and Np. The sediments are generally not affected with respect to their sorption behavior. To maintain realistic conditions the experiments were often carried out in an inert gas atmosphere consisting of 99 % Ar and 1 % CO_2;

Aerobic / anaerobic conditions
Eh und pH value
Grain size
Volume to mass ratio
Concentration of complexing agents • technical (EDTA, tensides, citric acid) • natural (humic acids, fulvic acids)
Colloid formation
Radionuclide concentration
Duration of the experiment
Solubility of the elements
Temperature
Radiolysis
Salinity of the groundwaters

Table 4: Investigated parameters which influence the sorption / desorption behavior of radionuclides

- furthermore, the sorption behavior of many elements is strongly influenced by the complexing agent content of the waters of natural origin (e. g. humic or fulvic acids) as well as technical origin (e. g. EDTA). In comparison with these substances, other complexing agents which may be present in the wastes, such as citric acid, are of minor importance;

- for some radionuclides a strong decrease in the sorption coefficient with increasing salinity of the ground waters is observed. This can be partially attributed to the increasing competition of chemically similar ions, e. g. in the case of Cs, or to the decrease in filterable colloids with increasing salinity, as in the case of actinides;

- the dependence of the sorption / desorption data on the pH value of the waters could only be estimated in a limited range. The reason is the high buffer capacity of the sediments which slowly corrects forced changes of the pH value in the direction of the natural system-specific ones.

To illustrate the sorption behavior of radiologically relevant radionuclides in natural systems, table 5 gives a survey of measured sorption coefficients in the overlying strata of the Gorleben salt dome. In general, the R_S values of systems with cohesive sediments are higher than in systems with sands which usually have only a low sorption capacity. The previously mentioned effect of the salinity of the waters can be seen in the case of Sr, Cs and Ra. The relatively low R_S values of Th are probably due to a complex formation of this element by humic substances which are found in Gorleben waters in varying amounts. These substances in combination with a possible microbial activity perhaps play a role in the enhanced sorption of iodine. In other systems a retention of iodine caused by microorganisms has been observed /3/.

Element	Sandy sediments		Cohesive sediments	
	Fresh water	Saline water	Fresh water	Saline water
Sr	1,8	0,3	8	1
Tc	1	1	1	1
I	1,5	0,6	38	5
Cs	44	1,2	984	105
Pb	262	772	10^4	$5\cdot10^4$
Ra	61	5	171	20
Th	360	150	670	870
Np	12	7,5	72	32
U	27	1	17	$14-1,4\cdot10^3$
Pu	10^3	10^3	$> 5\cdot10^3$	$> 5\cdot10^3$
Ac	80	100	$3,2\cdot10^3$	$2,3\cdot10^3$
C	10	1,6	3,1	2,7

Table 5: Sorption coefficients R_S (cm^3/g) in natural systems (Gorleben site). As of: 12/86

For Np and to a lesser extent also for Tc, relatively high sorption values have been measured in Gorleben systems. It can be assumed that reduced forms are involved in the sorption processes of these ions.

Table 6 contains the measured R_S values of relevant radionuclides for the Oxfordium, a formation belonging to the Malm stratigraphic unit in which the Konrad repository will be constructed.

Element	10^{-3} MOL/L EDTA	10^{-4} MOL/L EDTA	10^{-5} MOL/L EDTA
Tc	0,7	0,7	0,7
Cs	30	30	30
J	0	0	0
C	0,2	0,2	0,2
Sr	0	0	0
Np	30	30	30
Pu	500	1000	1000
Pb	0,6	6	40
U	1,8	2	3
Ra	1,6	1,6	1,6
Th	70	70	1000
Ac	30	70	200

Table 6: Sorption coefficients R_S (cm^3/g) for the Oxfordium (Konrad site). As of: 9/86

The values have been measured in the corresponding formation water which contained different amounts of the technical complexing agent EDTA (Ethylene-diamine-tetra-acetate). EDTA is often used for decontamination purposes in nuclear facilities and may therefore be a component of radioactive wastes. EDTA is released together with the radionuclides from the waste packages in the anticipated waterpath scenarios. With increasing distance from the repository area its influence on the sorption behavior of the radionuclide diminishes due to either radiolytical decomposition or dilution.

As can be seen from the data in table 6, the sorption of Pu, Pb, Th and Ac is considerably lower in waters containing EDTA, and this must be taken into account in modelling the radionuclide migration. In Gorleben systems, where in contrast to Konrad groundwaters humic substances have been found, the influence of these natural complexing agents has also to be taken into account.

The increase in the R_S values of U in table 6 is not due to the influence of EDTA, but to the strong dependence of the uranium sorption on the concentration. With increasing dilution leading also to a decrease in the EDTA concentration, the concentration of U decreases and therefore its sorption coefficient increases.

4 CONCLUDING REMARKS

As has been illustrated above, the results of the extensive sorption / desorption experiments have enabled the PTB to provide a realistic data base for model calculations concerning radionuclide migration through the geosphere. As a result, in the safety analysis for the post-operational phase of the planned repositories, the retardation capability of the host rock and the overlying strata can be characterized by sorption data which reflect the natural conditions.

LITERATURE

/1/ "Sicherheitskriterien für die Endlagerung radioaktiver Abfälle in einem Bergwerk".
Bundesanzeiger, 35, 5. Januar 1983, 4

/2/ Warnecke, E.; Tittel, G.; Brennecke, P.;
Stier-Friedland, G.; Hollmann, A.:
"Experimental Investigations of Possible Radionuclide Releases from the Planned Repositories in the Gorleben Salt Dome and Konrad Iron Ore Mine as Part of the Long Term Safety Assessment".
Proc. Int. Symp. on the Siting, Design and Construction of Underground Repositories for Radioactive Wastes, Hannover 3-7 March 1986, IAEA Vienna, 1986, 401

/3/ Behrens, H.:
"Zur Übertragbarkeit von Labor-Sorptionsdaten des Radiojods auf In-Situ-Verhältnisse".
Proc. Sem. Chemie und Migrationsverhalten der Aktinoide und Spaltprodukte in natürlichen aquatischen Systemen, (Kim, J.I.; Warnecke, E.; eds.) Report PTB-SE-14, Physikalisch-Technische Bundesanstalt Braunschweig, 1986, 184

Measurement of characteristic package data within the framework of the "ILW and HTGR fuel element test disposal in boreholes" project

K. Kroth, E. Barnert, D. Niephaus, H.-G. Götze

SUMMARY

The Federal Ministry for Research and Technology is sponsoring a project to develop and test the so-called borehole technique for the final disposal of certain types of intermediate level waste (ILW) and spent fuel elements from high-temperature gas-cooled reactors (HTGR).

The general objective of the project is to develop a technique for the final disposal of 1) fuel hardware, 2) cladding hulls and 3) dissolver sludges from the reprocessing of LWR fuel elements and of 4) spent HTGR fuel elements in boreholes in a salt repository. This technique will be demonstrated in a fully active but retrievable emplacement test in the abandoned Asse salt mine.

For this test, six specially instrumented ILW waste packages (standard 200 l drums with removable inserts in which the ILW is cemented in place) have been produced by the Karlsruhe Reprocessing Plant. For the emplacement test of HTGR fuel elements, four stainless steel cans filled with spent fuel elements from the Jülich High-Temperature Reactor (AVR) are available at KFA.

The basic nuclear data (total activity, dose rate and nuclear heat production) of the waste packages used in the emplacement test were determined experimentally. The results are compared with burn-up calculations.

The concentration of radioactive and non-radioactive gases in the boreholes may affect the radiological and operational safety of the repository. In order to quantify these concentrations, the release rates of H-3, C-14, Kr-85 and radiolytically generated hydrogen from the inserts are measured in a hot cell experiment.

Dr. K. Kroth, Dr. E. Barnert, Dr. D. Niephaus, Kernforschungsanlage Jülich GmbH, D-5170 Jülich, Federal Republic of Germany
H.-G. Götze, Wissenschaftlich-Technische Ingenieurberatung GmbH, D-5177 Titz-Rödingen, Federal Republic of Germany

INTRODUCTION

At the Kernforschunsanlage Jülich, KFA, (Nuclear Research Center Jülich) a project is being carried out on "Further Development of the Borehole Technology for the Disposal of Radioactive Wastes in Salt, with the Examples of Dissolver Sludge, Fuel Element Claddings, Fuel Hardware and HTGR Fuel Elements". The project is sponsored by the Bundesminister für Forschung und Technologie, BMFT, (Federal Ministry of Research and Technology).

The main objective of the project is to develop a technique for the final disposal of heat-generating intermediate level waste (ILW) and spent fuel elements form high-temperature gas-cooled reactors (HTGR) in unlined vertical boreholes in salt, and to test this technique under real conditions in a salt mine. In order to achieve this objective the project is divided in two subprojects with the following working items (Fig. 1):

Main objective of the R & D project: Development of a technique for the final disposal of dissolver sludge, cladding hulls and spent HTR fuel elements in unlined boreholes in salt, and testing of this technique in the Asse salt mine	
Sub-project I: **Emplacement and borehole sealing techniques**	**Sub-project II:** **Retrievable emplacement test**
● Development of components for the final repository	● Determination of gas release from the waste packages
● Safety engineering analyses	● Studying the interactions between salt and waste packages
● Cold testing (i.e. without real waste) of the prototypical components in the Asse salt mine	● Retrievable emplacement test with real waste packages in the Asse salt mine

Fig. 1: Objectives of the project "ILW and HTGR Fuel Element Test Disposal in Boreholes"

Subproject I: Emplacement and Borehole Sealing Techniques

The first working item is to develop all mechanical components necessary for the emplacement of 400 l waste packages. Main components are: shielded transfer cask, emplacement machine and borehole shielding slide. Additionally the plug for the final sealing of the borehole is to be developed. The prototype components will be tested in an inactive in-situ test programme in the Asse salt mine with respect to their possible use in the planned repository at Gorleben.

Subproject II: Retrievable Emplacement Test

In subproject II the major emphasis is placed on the experimental demonstration of the safe underground handling of real radioactive 200 l waste packages as part of an active emplacement test in the Asse salt mine. Due to the licensing situation of the Asse salt mine, the active waste packages have to be removed at the end of the test.

The release of corrosive gases from the rock salt into the borehole atmosphere under the action of radiation and/or temperature is studied during the disposal test.

The borehole atmosphere will be contaminated with radioactive gases and radiolytically generated hydrogen leaking from the waste packages. In order to quantify the expected concentrations of gaseous activity and hydrogen in the atmosphere of the boreholes, the release rates of these components from the waste packages to be used in the emplacement test are measured during a hot cell experiment at KFA.
The subject of this paper is limited to the last working item.

2 WASTE PACKAGES STUDIED

Within the framework of the emplacement test in the Asse salt mine, six waste packages with heat-generating ILW from a reprocessing plant will be experimentally emplaced. The gas release from these packages is being studied in a hot cell experiment now underway.

The waste packages were produced and cemented in place at the Karlsruhe Reprocessing Plant (WAK) according to the usual practice there. The waste originates from three fuel elements from the Neckar Westheim power station (GKN). The data and irradiation history of the three fuel elements are compiled in table 1.

nuclear reactor	GKN, PWR
initial enrichment	3.2 %
burn-up	37.5 GWd/tU
end of irradiation	7/82
date of reprocessing	5/86
fuel mass per element	357 kg
no. of fuel elements	182, 187, 190

Table 1: Burn-up data of the reprocessed GKN-fuel elements

The wastes are packed in 200 l drums sealed by a double-lid system. The outer lid has an angle ring with the aid of which the drums can be manipulated by remote control with an appropriate grapple. So-called 120 l removable inserts are located inside the drums and serve to accommodate the wastes. The removable inserts were instrumented with Pt-100 temperature sensors before being filled with waste. These sensors are used to measure the product temperatures during the gas release experiment.

The wastes placed into the removable inserts were cemented in place, according to normal practice at the WAK. Portland cement, PZ 45F, with a water/cement ratio of 0.45, was used.

Approx. 1 % of concrete thinner was added to thin the cement paste. The following wastes were conditioned in the removable inserts of the six ILW drums:

2.1 FUEL HARDWARE (FH-187)

One of the six 200 l drums contains the fuel hardware of the GKN fuel element no. 187. The fuel hardware of a GKN element comprises the end pieces, the control rod guide tubes and six grid spacers. The fuel hardware has a weight of about 50 kg. It is made from stainless steels of different composition and from Inconel. These materials are activated by (n,γ)-reactions in the neutron flux of the reactor. The major source of activity is Co-60 generated by (n,γ)-reactions on Co impurities in the fuel hardware materials. Fig. 2 shows the nuclear data from the fuel hardware drum. The values specified for overall activitiy, surface dose rate, and nuclear heat output - indicated by 1) in Fig. 2 - are the results of an ORIGEN burn-up calculation. The activation calculation is based on the known chemical compositions of the fuel hardware and certain assumptions about additional contaminations of these materials. In the activation calculations, the vertical flux distribution over the reactor core was taken into account. There is relatively good agreement between the experimentally determined γ-dose rate and the forecast from the ORIGEN burn-up calculation, if it is remembered that the gamma dose rate is largely governed by the extent of Co-contamination of the fuel hardware materials. However, this is largely unknown and is only available in the form of an upper specification value for the various materials.

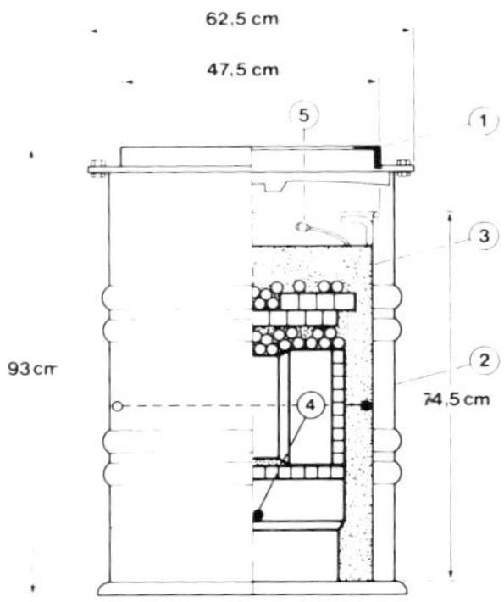

(1) lid with lifting ring
(2) 200l drum
(3) removable insert
(4) resistance thermometer
(5) cable connector

contents:	50 kg	fuel hardware
mass:	320	kg
total activity:	$1.3 \cdot 10^{14}$	Bq[1)]
surface dose rate:	90	Gy/h[1)]
nuclear heat production:	16	Watt[1)]
γ-dose rate; 0.1 m:	30	Gy/h[2)]

[1)] ORIGEN burnup calculation;
4.5 years after discharge from the reactor
[2)] experimental results; 4.3 years after discharge from the reactor

Fig. 2: 200 l standard drum with fuel hardware cemented in place for the retrievable emplacement test

2.2 CLADDING HULLS (CH-187, 190)

Two of the six ILW drums contain leached cladding hulls. Approx. 60 % = 60 kg of the cladding hulls from fuel elements no. 187 and no. 190 are contained in each of the two drums. The cladding hulls are manufactured from Zircaloy-4. Apart from zirconium, the main constituent of Zircaloy-4 is tin. The major activation products are Co-60 and Sb-125. The cladding hulls are moreover contaminated with fission products and α-active actinides.

The nuclear data of the two cladding hull drums are compiled in Fig. 3. The values specified for the overall activity, surface dose rate, and nuclear heat output -indicated by 1) in the Figure- are the results of an ORIGEN burn-up calculation. The activation calculation is based on the known composition of the Zircaloy-4 and takes into consideration the contamination of the cladding hulls with fission products and actinides.

The gamma-dose rate and nuclear heat output of the two cladding hull drums were experimentally determined at KFA. The results are indicated in Fig. 3 by 2.

The nuclear heat output of the cladding hull drums was determined by an isothermal heat flux calorimeter, which was developed by the Technical Engineering Department (IT) at the Karlsruhe Nuclear Research Centre. The calorimeter was made available to KFA for these experiments.

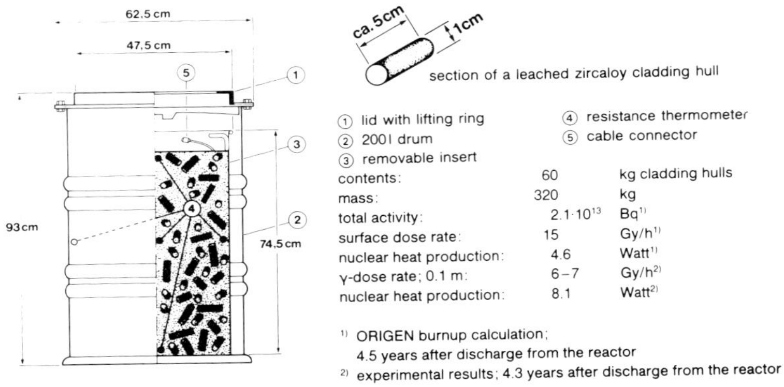

Fig. 3: 200 l standard drum with cladding hulls cemented in place for the retrievable emplacement test

The experimental results for the gamm-dose rate and nuclear heat output are in good agreement with the forecast of the ORIGEN activation calculation. This agreement is remarkable since the degree of Co impurity in the cladding hull material and the extent of contamination of the cladding hulls by fission products are of decisive significance for both values, and these factors involve a great deal of uncertainty.

2.3 DISSOLVER SLUDGE (DS-182, 187, 190)

An insoluble residue, termed dissolver sludge (DS), remains after the dissolution of irradiated nuclear fuels. The dissolver sludge includes phase precipitates containing noble metals, zirconium chips, and possibly undissolved fuel particles. The dissolver sludge must be quantitatively separated from the fuel solution as far as possible before the latter is fed into the chemical process section of the reprocessing plant.

The fuel solution is purified at the WAK by using a coarse-meshed polypropylene filter bag through which the fuel solution is filtered off. It is well known that this type of feed filtering is not suitable for quantitatively separating the fines of the dissolver sludge from the fuel solution. This operation requires finer filter materials with pore sizes of $\emptyset \leq 1\ \mu m$.

The WAK provided three 200 l waste packages with dissolver sludge cemented inhomogeneously in place for the emplacement test at the Asse salt mine. Each package contains two of the filter bags mentioned above through which the fuel solution of one GKN fuel element was filtered. The γ-dose rate, nuclear heat output, and overall activity of the three dissolver sludge drums were determined experimentally at KFA.

The nuclear data from the three dissolver sludge waste packages are compiled in Fig. 4. The data for the overall activity and the nuclear heat output indicated by 1) correspond to the DWK/PWA (DWK - Deutsche Gesellschaft für Wiederaufarbeitung von Kernbrenn stoffen mbH; PWA - Projekt Wiederaufarbeitung und Abfallbehandlung, KfK, Karlsruhe) reference proposal for the nuclear composition of dissolver sludge /1/. In order to render these data comparable to the experimentally determined package data, the reference data were converted to the fuel mass of a GKN fuel element and to a decay period of 4.5 years.

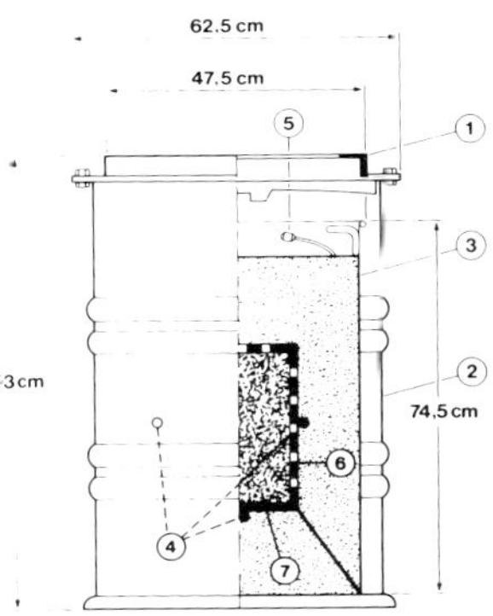

1. lid with lifting ring
2. 200 l drum
3. removable insert
4. resistance thermometer
5. cable connector
6. box
7. filter bag with dissolver sludge

contents:	dissolver sludge from a GKN fuel assembly according to the WAK technique	
mass:	320	kg
total activity:	$3.2 \cdot 10^{14}$	Bq[1)]
nuclear heat production:	36	Watt[1)]
γ-dose rate; 0,1 m:	0.3–0.55	Gy/h[2)]
nuclear heat production:	1.5–2.7	Watt[2)]
total activity:	$5.4–9.7 \cdot 10^{12}$	Bq[2)]

1) according to the DWK-PWA reference: 4.5 years after discharge from the reactor

2) experimental value; 4.3 years after discharge from the reactor

Fig. 4: 200 l standard drum with dissolver sludge cemented inhomogeneously in place for the retrievable emplacement test

A comparison of the experimentally determined heat outputs with those from the DWK/PWA reference proposal indicates that the nuclear inventory of the dissolver sludge waste packages is lower by a factor of 13-24 in relation to the expected value of the reference proposal.

A comparison of the total activities provides an even greater discrepancy.

The deficit found for nuclear heat output and total activity has to be interpreted in such a way that the feed filtering process practised at the WAK is not suitable for quantitatively separating the particularly active fine-grained fraction of the dissolver sludge. This results in an important consequence for the emplacement test, namely that the packages available cannot be regarded as representative of the dissolver sludge packages that are expected from the Wackersdorf reprocessing plant.

3 PRECHARACTERIZATION OF THE ILW WASTE DRUMS

After delivery of the waste drums to KFA, the packages were subjected to a precharacterization /2/. The aim of this examination was to experimentally determine the activities of H-3 and Kr-85 released into the drum gas volume during interim storage. Additionally the amount of radiolytically generated hydrogen is to be determined. At the time of the examination, the 200 l drums had been sealed gas-tight for a period of 96-131 days. During this period of interim storage, the gaseous radionuclides released from the cement stone matrix and the released hydrogen accumulated in the free gas volume of the 200 l drums. The free gas volume is approx. 89 l.

The precharacterization was carried out in Cell 505 of the Large Hot Cells (GHZ) at KFA. Two holes were drilled in the drum lids with the aid of a remote-controlled drilling device. Two gas sampling probes were placed into the boreholes connected via hose pipes to a gas analysis apparatus located outside the cell. Precautions were taken during this so-called "puncturing" of the drums to ensure that a mixing of the drum gas with the cell air remained negligible. Fig. 5 shows a schematic representation of the experimental assembly.

After establishing the connection lines, the drum gas was first homogenized by recirculation. The moistre entrained in the gas flow was thus precipitated in the freezing vessel. The H-3 activity of the water frozen out was analyzed with the aid of liquid scintillation after purification by distillation. The Kr-85 activity concentration of the drum gas was determinded by γ-spectrometry using a high-resoluton Ge-detector. To this end, the pressure vessel (7) was loaded with drum gas up to a pressure of 6 bar. The Kr-85 was quantitatively

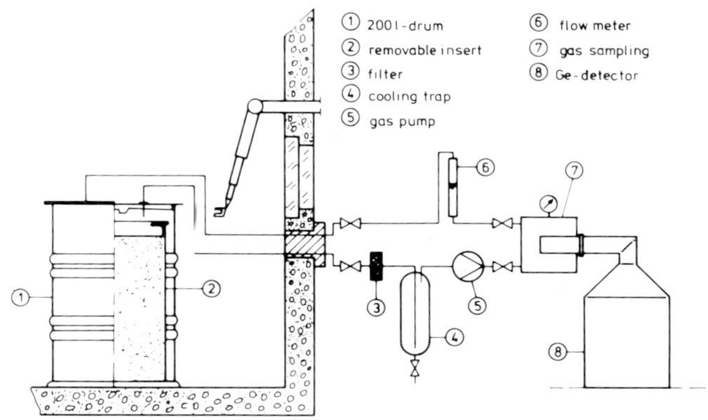

Fig. 5: Experimental setup for the precharacterization of the 200 l waste drums

determined by an evaluation of the 514 keV line in the recorded γ-spectra. The drum gas drawn off into the pressure vessel served as a gas sample for an analysis of the gas constituents by gas chromatography.

With the exception of the fuel hardware drum, it was observed that the internal drum pressure was lower than the ambient pressure. During interim storage, the drum containing fuel hardware had become leaky probably due to overpressure. Since an unknown fraction of the drum gas had apparently escaped, gas analysis had to bee omitted in this case.

The results of the precharacterization of the six ILW waste drums are shown in Tab. 2. These results will be interpreted in a later section.

4 DESCRIPTION OF THE HOT CELL EXPERIMENT

The experimental assembly basically consists of six overpacks, each of which accommodates a removable insert with cemented wastes. In order to be able to reliably detect the released gaseous components the overpacks are designed in such a way that their tightness is better than 10^{-5} mbar·l/sec. The overpacks are designed as pressure vessels up to a pressure of 5 bar. The overpack lid is equipped with a rupture

MAW	fuel element no.	A_{Kr-85} mCi	A_{H-3} nCi/ml	gas components N_2 %	O_2 %	H_2 %	D_{LY} 0.1 m R/h [1]	D_{LY} 1.0 m R/h [1]
fuel hardware	187	[2]	–	–	–	–	3000	250
cladding hulls	187	10	67	93	7.4	1.7	600	52
cladding hulls	190	5.8	42	94	4.6	0.53	700	55
dissolver sludge	187	0.77	16	85	8.4	5.0	30	1.7
dissolver sludge	190	2.3	6	79	5.2	9.8	55	3.5
dissolver sludge	182	1.56	0.3	84	6.2	5.8	38	3.5

[1] distance from the drums surface

[2] not determined since drum was leaking

Table 2: Gas release from the waste matrices into the voids 3-4 months after sealing the drums

disk which blows off at an overpressure of 3 bar. There is a container sump at the bottom of the overpack in which condensed moisture is collected. The condensate can be withdrawn via a suction line and then analyzed. The overpack lid is equipped with a pressure gauge to monitor the internal container pressure; the overpack lid also contains the electrical penetrations to connect the temperature sensors to the data acquisition system.

In order to be able to study the gas release at elevated repository-relevant temperatures (up to max. 90° C), the overpack is equipped with a jacket heating with which the entire overpack and its contents can bee heated.

The overpacks are accommodated in the Hot Cell due to the intensive radiation emanating from the waste packages. The appropriate technical boundary conditions thus had to be established for the delivery and unloading of the transport casks and the remote-controlled loading of the overpacks.

Each of the six everpacks is connected to a valve-control panel locatec outside the Hot Cell via two tubes (a feed line and a return line). Each of the six overpacks can be connected via the valve-control panel to various devices for gas analysis.

The Hot Cell experiment is shown diagrammatically in Fig. 6.
Fig. 7 shows the arrangement of the six overpacks in the Hot Cell.

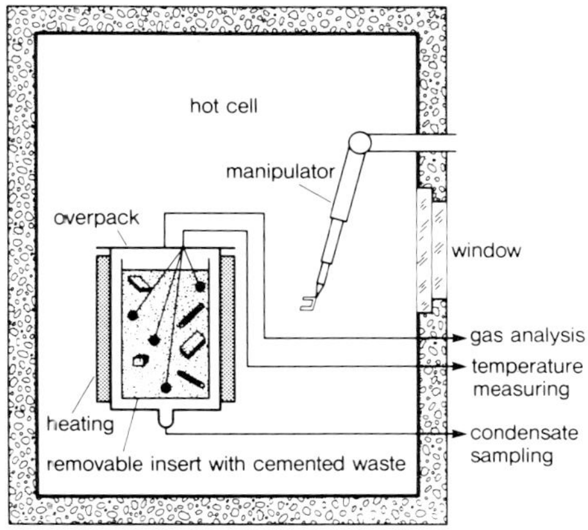

Fig. 6: Hot cell experiment to determine the gas release (H_2, H-3, Kr-85) from ILW waste packages

5 FIRST RESULTS FROM THE HOT CELL EXPERIMENT

In late 1986, five of the six overpacks were loaded with the corresponding removable inserts. After sealing the overpacks, the free volume of the overpacks was filled with synthetic air at a pressure of 1012 mbar. The empty volume is approx. 45 l. After a closure time of approx. two months, gas samples were taken from the overpacks which were examined with respect to their Kr-85 activity and to their gas composition. The temperature of the waste packages was 23° C. The internal pressure in the overpacks was recorded during the closure time.

The results of the pressure measurements will first be represented and interpreted.

Figs. 8 + 9 + 10 show the experimentally determined pressure variation in the overpacks FH-187, CH-190 and DS-187. The dotted lines represent the variation in pressure. After being sealed for 40-60 days, gas samples were taken from the overpacks. The results of the gas analysis are also represented in Figs. 8 + 9 + 10. The pressure in the overpacks drops by approx. 100-150 mbar due to gas sampling. After the gas analysis, the gas samples were fed back into the corresponding containers. The pressure then practically rises again to the initial value.

Fig. 7: Six overpacks loaded with removable inserts installed in a hot cell

A simple phenomenological model describing the influence on pressure due to changes in the gas composition has been developed. The model is based on the following assumptions:

1) The N_2 partial pressure of 790 mbar arising from the original filling of the container volume with air remains constant in time.

2) Radiolytically generated hydrogen results from the interaction of β/γ-radition with the free pore water in the cement stone. The H_2 production rate is regarded as constant for the periods of interest here. This assumption leads to a linear rise of the H_2 partial pressure, pH_2.

3) The oxygen enclosed in the overpacks together with the air is consumed by radiation-induced oxidation processes (ozone formation). The initial O_2 partial pressure is 212 mbar.

On the basis of these assumptions, the overall pressure can be represented as the sum of three partial pressures as follows:

$$P_{\Sigma} = P_{N_2} + \dot{P}_{H_2} \cdot t + P_{O_2}(0) \cdot \exp(-\lambda_{O_2} \cdot t) \qquad (1)$$

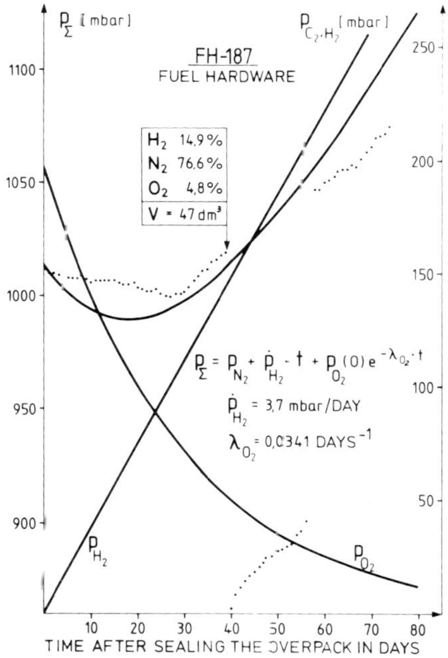

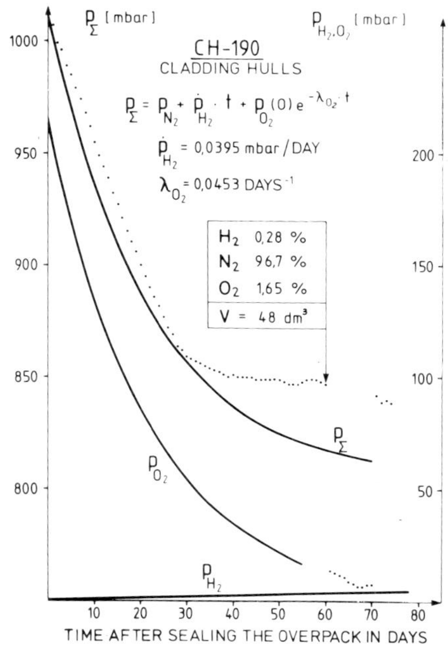

Fig. 8: Evaluation of the gas pressure in the overpack loaded with fuel hardware

Fig. 9: Evaluation of the gas pressure in the overpack loaded with cladding hulls

PH_2 is the rate of pressure increase as a consequence of the formation of radiolytically generated hydrogen; λO_2 is the time constant with which the O_2 partial pressure exponentially decreases.

The two free parameters in Eq. 1 may be calculated using the analytical data for the composition of the drum gas. The results are shown in Figs. 8 + 9 + 10, respectively.

The overall pressure P_{Σ} and the partial pressures PH_2 and PO_2 can then be calculated as a function of time using these data. The overall pressure and the partial pressures thus calculated are additionally shown in Figs. 8 + 9 + 10.

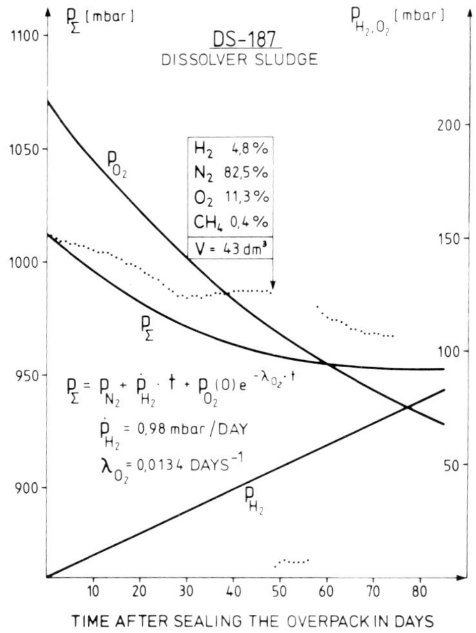

Fig. 10: Evaluation of the gas pressure in the overpack loaded with dissolver sludge

Qualitative predictions can be made about the medium-term pressure variation in the overpacks and in the 200 l drums, respectively, on the basis of this model:

1) FH-187 (Fig. 8):
   After a brief starting-up phase (approx. 30 days), a linear rise in pressure at a rate of approx. 3.7 mbar/day occurs.

2) CH-190 (Fig. 9):
Exponential pressure decrease down to approx. 850 mbar; the initial pressure will be reached after a few years.

3) Ds-187 (Fig. 10):
Slight pressure decrease over a period of approx. 100 days can be seen; thereafter a linear pressure rise at a rate of approx. 1 mbar/day will occur.

The data relate to a free volume in the overpack of approx. 45 l. In the case of larger free volumes, e. g. in the standard 200 l drum, the pressure changes will be correspondingly slower.

If it is assumed that the release rates for H_2 and Kr-85 remain constant within the measuring period from December 86 to March 87 then the corresponding release rates can bee calculated from the integral data. Tab. 3 gives a summary of the H_2 and Kr-85 release rates calculated in this way. The results partly originate from drum measurements and partly from measurements in overpacks. The measured values indicated by 1) are averages over two measuring periods.

waste package	hydrogen release rate ml/day	Kr-85 release rate µCi/day
FH-187	183	0
CH-187	13	83
CH-190	3[1)]	43[1)]
DS-187	34[1)]	4,4[1)]
DS-190	74[1)]	11[1)]
DS-182	46	14

Table 3: Release rates of hydrogen and Kr-85 from the ILW waste packages

1) mean value from the results of two consecutive measuring periods

As expected, no Kr-85 was released from the drum insert with fuel hardware. With respect to the release of Kr-85 from the remaining packages, there are clear indications that the release rates need not be assumed to be constant over long periods but rather that they decrease with time. As yet there are no indications for a corresponding decrease in the H_2 release rates.

It is well known that leached Zircaloy cladding hulls contain Kr-85 in an internal corrosion layer a few um in thickness. The Kr-85 atoms present in this layer result from fissions having taken place in the outer boundary zone of the fuel. In this way, among all other fission products, Kr-85 atoms are recoil implanted into the inner surface of the claddings. A corrosive attack by the highly alkaline cement paste on the inner oxide layers of the cladding hulls seems to us to be the most plausible explanation for the relatively high Kr-85 releases into the drum gas.

Apart from precipitates from noble metals, dissolver sludge also contains zirconium chips and abraded particles produced during shearing of the fuel rods. In the case of the dissolver sludge packages it is conceivable that these materials are partly "decomposed" by the cement paste involving a release of the bound Kr-85.

LITERATURE

/1/ P.-G. Maurer, Systemanalyse zur Untersuchung und Beurteilung von alternativen Verfestigungsmatrices und -Verfahren für Feedklärschlamm, Nukem Bericht FuE 80050 (1980)

/2/ K. Kroth, Untersuchung zur Freisetzung flüchtiger Radionuklide und Radiolysegase, in: MAW und HTR-BE Versuchseinlagerung in Bohrlöchern, Jül-Conf-60 (1987)

Safety analyses for the operational phase of the planned KONRAD repository

H. Illi

SUMMARY

The safety analyses of the operational phase for the planned federal Konrad repository are described. With the help of these analyses it is being checked whether the radiological effects of the repository on man and surroundings are in accord with the protection goals defined in the Atomic Energy Act and the Radiation Protection Ordinance. Starting from the potential exposure of the operating staff and the surroundings of the repository which was determined on the basis of models, requirements have been iteratively derived for the design of the repository, its operation and the radwaste, until the protection goals were complied with.

Dr. H. Illi, Physikalisch-Technische Bundesanstalt, Bundesallee 100, Postfach 33 45, D-3300 Braunschweig, Federal Republic of Germany

1 INTRODUCTION

During the period of operation and the postoperational phase of a repository, protection goals defined by the Atomic Energy Act (Atomgesetz, AtG) /1/ and the Radiation Protection Ordinance (Strahlenschutzverordnung, StrlSchV) /2/ must be complied with. The observance of the protection goals is checked and proved with the aid of safety analyses. Taking the radiological effects on the operating staff and the environment of the repository into account, which are determined on the basis of model assumptions, in iterative steps, requirements are derived for the planning and construction of the plant and the wastes to be disposed of until observance of the protection goals is guaranteed.

For the planned Konrad repository, safety analyses have been carried out with regard to normal operation, incidents, the thermal influence upon the host rock and with respect to criticality and long-term safety; requirements for the design of the repository, its operation and the waste forms, packagings and the activities of radionuclide mixtures in waste packages have been derived. The latter are limited by activity limiting criteria which result form different fields for which requirements must be defined and which exist independent of one another.

From the point of view of safety engineering, the volatility of radioactive substances is of importance in normal operation; in the case of incidents, the formation of radioactive aerosols and gases is a decisive factor. The heat output of the waste packages plays an important role as far as the thermal influence upon the host rock is concerned, and the critical mass of the radionuclides is of importance with regard to criticality safety.

The way of proceeding and the results of the safety analyses for the planned Konrad repository will be treated in the following.

The report is based, among other things, on the work carried out on behalf of the PTB by the Gesellschaft für Reaktorsicherheit mbH (Company for Reactor Safety), Cologne.

2 WASTES TO BE DISPOSED OF

It is intended to emplace into the planned federal Konrad repository radioactive wastes with negligible or low heat generation coming from the reprocessing of spent fuel elements, from nuclear power plants, research establishments, the collecting depots of the federal states, from nuclear industry, the decommissioning of nuclear facilities and from other origins.

The mine workings of the Konrad mine allow storage cavities of about 1 million cubic metres volume to be driven. Depending upon the waste package volume to be disposed of annually - up to 3400 transport units in one-shift operation and twice the volume in two-shift operation - a period of operation of about 40 years results if the cavities are made use of by 50 %; these figures are based on prognoses of conditioned radioactive waste arising in future which were made in 1986 /3/. The waste spectrum investigated in the safety analyses was determined on the basis of previous evaluations made in 1984, in which a waste production of 0,5 million cubic metres of conditioned radioactive waste was assumed. Fig. 1 shows the portions in the volume of the produced waste as a function of its origin.

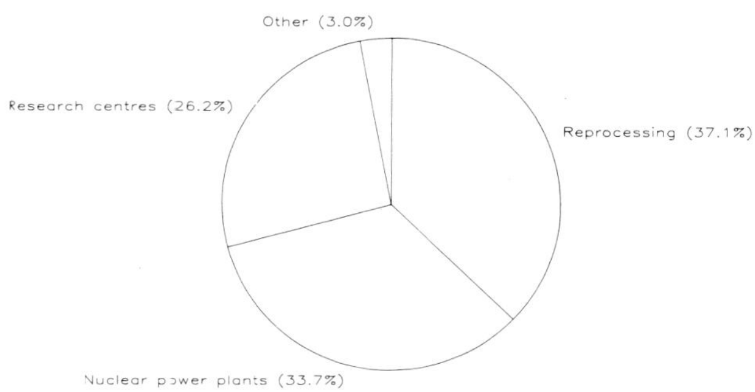

Fig. 1: Portions in the volume of the waste spectrum taken as a basis for the safety analyses; total waste volume: 500 000 m^3

The determination of the waste quantities was based on the following assumptions:

- For an assumed extension scheme, the power of the nuclear power plants installed in the year 2000 will amount to about 28 GW (17 pressurized water reactors, 5 boiling water reactors);

- 2800 t of spent fuel elements originating from light-water reactors will be reprocessed in the Federal Republic until the year 2000; from 1990, radioactive waste resulting from German nuclear fuel reprocessed abroad will be returned;

- The amount of radioactive waste arising annually in the collecting depots of the federal states, large research centres, nuclear industry and other origins is assumed to remain unchanged until the year 2000; its amount will correspond to the waste volume determined for the year 1984.

- Waste arising from the decommissioning of nuclear facilities over a period of 5 years (from 1989 to 1993) from the following plants: Karlsruhe research reactor 2 (FR-2), Großwelzheim superheated steam reactor (HDR), nuclear system of the nuclear-propelled research craft "Otto Hahn", Niederaichbach nuclear power plant (KKN), Kahl nuclear power test plant (VAK), Jülich nuclear power test plant (AVR).

The waste spectrum based on the above assumptions covers 165 different waste types and 150 radionuclides. As the radioactive wastes are not handed over to a repository immediately after their production, only radionuclides with half-life periods of more than 10 days have been taken into account in the incident analyses. This restrictive criterion has also been applied in the analyses of the thermal influence upon the host rock, an exception being made of actinides. On the basis of this cut-of-criterion, 80 radionuclides have been investigated within the scope of incident analyses, and 106 radionuclides in the analyses made with respect to the thermal influence upon the host rock. Six radionuclides of radionuclide groups which must be classified as volatile have been dealt with in the safety analyses for normal operation.

3 SAFETY ANALYSES

The safety analyses have been carried out in accordance with the principles stated in the "Safety criteria for the disposal of radioactive waste in a repository" /4/, i. e. they take site-specific factors as well as the volume, kind and properties of the radioactive wastes to be disposed of into account. On the basis of the results of the safety analyses, requirements for the design of the repository, its operation and the wastes to be disposed of have been derived (which must be regarded as provisional waste acceptance requirements until the plans have been approved).

The provisional waste acceptance requirements were derived in two steps. First of all, the safety analyses were carried out with the model nuclide spectrum explained above, the volume and the properties of the radioactive wastes being assessed according to the respective technical safety aspects, e. g. their volatility, heat generation etc. These analyses gave a survey of the properties of today's wastes which are important from the viewpoint of safety, and defined the wastes which are to be considered for reposition. As it is well known that the nature and composition of radioactive wastes may change, also with respect to future developments in their conditioning, waste acceptance requirements had to be defined which took today's waste to a sufficient extent into account as far as waste conditioning and waste composition were concerned and which offered as great as possible a lexibility for the waste spectrum of future decades.

This objective has been tackled in a second step. Based on the evaluation of the model nuclide spectrum explained above, so-called quality characteristics were defined for waste forms (processed radioactive waste without packaging) and for the packagings of the wastes (containers and boxes) and activity limiting criteria were established for the activity of radionuclide mixtures which - harmonized with, and supplemented by, requirements for the operation and design of the planned repository- guarantee that the plant complies with the protection goals laid down in the Atomic Energy Act and the Radiation Protection Ordinance. The quality characteristics which comprise general basic requirements valid for all wastes, and special requirements for individual waste form groups and packagings, represent, together with the activity limiting criteria, the provisional waste acceptance requirements; there is no longer any direct relation to the model nuclide spectrum.

The basic requirements which are valid for all wastes comprise, for example, the exclusion of fermenting and rotting materials, of gases and liquids (except for reasonably attainable and unavoidable residual contents) in ampoules, bottles and other containers. They furthermore cover qualitative requirements for the setting and solidification of immobilization materials and their being cast into cavities, and of wastes. Quantitative requirements exist for the volume of waste packages (limitation to 20 t), the mass per unit volume of fissionable products in the waste form, a concrete shielding of wastes which may release Rn-220, the local dose rate of waste packages (on their surface and at a distance of 1 or 2 m from the surface) and their surface contamination. Standardized containers (cylindrical packagings made of concrete or cast material and boxes of sheet metal, concrete or cast material) are prescribed for the packaging of the wastes. More details are given in /5/.

The quality criteria which are not comprised in the basic requirements result from various safety analyses. They are supplemented by activity limiting criteria which can be formulated in the form of the following summation criterion:

$$S = \sum_i \frac{A(i)}{G(i,p,k,c)} < 1. \tag{1}$$

where:

S Summation value

A(i) Activity of the radionuclide i/the radionuclide group i in a waste package (incident analyses, thermal influence upon the host rock, criticality safety) or the total activity of the radionuclide i/the radionuclide group i emplaced in one year of operation (normal operation).

G(i,p,k,c) Activity limiting value of the radionuclide i/the radionuclide group i, determined on the basis of safety analyses for waste form group p (incident analyses 6, normal operation 2) and class k of the waste containers (incident analyses 2, normal operation 4). In the case of normal operation, the activity limiting value refers to the total activity which can be emplaced in one year of operation; in this case, c specifies the chemical form of the radionuclide or other characteristic features, such as the mass fraction of water in the waste form, the HTO concentration in the water of the waste form or the total activity in the waste form without tritium activity. For the criticality safety and the thermal influence upon the host rock, only the geometric dimensions of the containers are of importance.

Waste form groups had to be defined (Table 1) in order to be able to describe the release of radionuclides from the most different waste forms within the scope of the various safety analyses. From the incident analyses, quality characteristics result for 6 waste form groups (Table 2). These waste form groups could be adopted without any change for the safety analyses for normal operation and long-term safety, a subdivision into only 2 groups being sufficient for the safety analyses for normal operation and a subdivision into 4 groups for the long-term safety analyses, an additional group referred to as "Others" being specified.

As the containers also have an influence upon the release of radionuclides from waste packages, quality characteristics had to be defined for the packagings of the wastes as well. In order to quantify the container barrier in the case of incidents, the containers were

Waste form group	Safety analyses
01: Bitumen and plastic products 02: Solid matter 03: Metallic solid matter 04: Compacted waste 05: Cemented/concreted waste 06: Concentrates	incident analyses
03: Metallic solid matter Others (01, 02, 04-06)	normal operation
01: Bitumen and plastic products 03: Metallic solid matter 05: Cemented/concreted waste Others (02, 04, 06)	long-term safety

Table 1: Waste form groups

subdivided into two classes and corresponding quality characteristics established (Table 3). For the safety analyses for normal operation, the leak tightness had to be defined as quality criterion for the packagings in the form of a specified annual release factor.

3.1 NORMAL OPERATION

The safety analyses for normal operation cover the exposure of the operating staff and the environment to radiation from the radiation fields emanating from the waste packages (direct and scattered radiation). In addition, the emission of volatile radionuclides released from the waste packages to the environment with the extracted air (from parts of the surface plant) and with the exhaust air (from the subsurface) were investigated as well as the inhalation of volatile radionuclides by the operating staff. Another subject of the investigation was the controlled introduction of liquid service and mine waters into surface waters.

From the wastes to be emplaced, tritium (as HTO and HT), carbon-14, iodine-129 and radon-222 (as decay product of Ra-226) may be released as well as solid matter (α- and β/γ-emitters) in the form of aerosols. The release rates of these radionuclides or radionuclide groups depend upon the waste form and the quality of the packaging. Corresponding model concepts are described in /6/ and /7/.

Activity release rates applied for have been derived for the above radionuclides (Table 4) in such a way that the exposure rate in the environment amounts to only about 10 % or 15 % of the limiting value for the whole body according to section 45 Radiation Protection Ordinance for the air- or water-borne activity discharge. The exposure of the operating staff to radiation as a result of the inhalation of volatile radionuclides originating from the wastes lies below 1 % of the dose limit for the whole body valid for persons of category A occupationally exposed to radiation (section 49 Radiation Protection Ordinance) (Table 5).

By "backward calculations", limiting values for the annually disposable activity have been derived from the activity release rates applied for for the above-mentioned volatile radionuclides (Table 6), taking into account retentions which are ensured in particular by so-called closing-off structures used to seal filled emplacement rooms.

Waste form group	Quality characteristics
01	- Basic requirements
02	- Basic requirements - Limitation of the activity portions of combustible wastes with melting points < 300 °C to ≤ 1 % in the waste form - Immobilization of the above-mentioned combustible material in such a way that it cannot emerge from the waste form in liquid state
03	- Basic requirements - Waste form is a metal or consists of reactor core material, with the exception of graphite
04	- Basic requirements - Waste form is compacted into an inherently stable form under a pressure of ≥ 30 MPa
05	- Basic requirements - Waste is immobilized in cement or concrete - Embedded or solidified waste, e.g. ashes, powders or aqueous concentrates, uniformly and completely distributed in the cement or concrete - Caste waste, e.g. scrap is distributed as uniformly as possible in the waste form - The compressive strength of the waste form is ≥ 10 N/mm^2 or the portion in the volume of waste matrix is > 40 %
06	- Basic requirements - The waste itself consists of a solid, non-powdery residue of evaporated liquids or dried sludges - The waste is not combustible

Table 2: Quality characteristics of the waste form groups, derived from incident analyses

Packagings	Quality characteristics
Class 1	- Basic requirements - Up to an impact velocity of ≤ 4m/s, the integrity of the packaging is preserved to such an extent that in the case of a subsequent thermal load, the access of oxygen to the waste form is so limited that flammable waste forms with melting points > 300 °C do not burn away with naked flame, but pyrolyse
Class 2	- Basic requirements - The packaging withstands a drop from a height of 5 m onto an unyielding target in such a way that the leak rate (relative to standard conditions such as in leak testing by the vacuum method) after the drop does not exceed 1×10^{-4} $Pa \cdot m^3/s$ (this does not apply to radioactive wastes immobilized in bitumen, when the waste is uniformly and completely distributed in the bitumen). - The wall of the packaging has a thermal conductivity resistance (thickness times reciprocal thermal conductivity) of ≥ 0.1 $m^2 \cdot K/W$ (lost concrete shieldings) or - the leak rate (relative to standard conditions such as in leak testing by the vacuum method) is < 1×10^{-5} $Pa \cdot m^3/s$. In the case of a fire with a temperature of 800 °C and a duration of one hour, the integral leakage of the gas released from the packaging during the fire and a cooling phase of 24 hours is < 1 mole.

Table 3: Quality characteristics of the packagings, derived from incident analyses

The following summation criterion is valid for the annually disposable activity of a volatile radionuclide:

$$S_b(i) = \sum_{p,k,c} \frac{A(i)}{G_b(i,p,k,c)} < 1 \qquad (2)$$

where:

S(i) Summation value (B = index for normal operation)

A(i) Activity of the radionuclide i/the radionuclide group i emplaced in the current year of operation

$G_B(i,p,k,c)$ Activity limiting value for the annually disposable activity of the radionuclide i/the radionuclide group i, determined from the safety assessment of normal operation for waste form group p, leak tightness class k and specification c.

Radionuclide/ radionuclide group	Activity discharges Bq/a	Activity discharges via
H 3	1.48×10^{13}	
C 14	3.7×10^{11}	
I 129	7.4×10^{6}	
Rn 222	1.85×10^{12}	
aerosols (half-life period > 10 d):		exhaust air
- α-emitters	3.7×10^{6}	
- β/γ-emitters	7.4×10^{7}	
H 3	7.4×10^{12}	
radionuclide mixture	7.4×10^{8}	waste water

Table 4: Release rates applied for the activity discharges with exhaust air and waste water

Section of the population	Exposure path	Dose equivalent per year (whole body) in 10^{-5} Sv/a
Staff Persons exposed to radiation (category A) section 49 Radiation Protection Ordinance	Inhalation of volatile radionuclides Direct and scattered radiation	< 50 on the average < 500
Environment section 45 Radiation Protection Ordinance	Exhaust air Waste water	2.3 3.8
Environment section 44/45 Radiation Protection Ordinance	Direct and scattered radiation	about 100

Table 5: Potential exposure to radiation in normal operation

Radionuclide/ Radionuclide group	Packaging without specified tightness		Packaging with specified tightness — Annual portion of release ≤0.01		≤0.001		≤0.0001	
	metallic solid matter	other waste form groups	metallic solid matter	other waste form groups	metallic solid matter	other waste form groups	metallic solid matter	other waste form groups
Tritium								
- unspecified		2.3×10^{13}		7.1×10^{13}		7.1×10^{13}		7.1×10^{13}
- as HTO*		a) 6.9×10^{13} b) 6.7×10^{13} c) 2.3×10^{13}		a) 6.8×10^{15} b) 2.4×10^{15} c) 7.1×10^{13}		a) 5.8×10^{16} b) 3.4×10^{15} c) 7.2×10^{13}		a) 2.4×10^{17} b) 3.6×10^{15} c) 7.2×10^{13}
as HT	4.7×10^{15}		9.6×10^{15}		9.6×10^{15}		9.6×10^{15}	
C 14								
- unspecified or in volatile form	5.3×10^{16}	1.0×10^{12}	9.6×10^{16}	1.9×10^{12}	9.6×10^{16}	1.9×10^{12}	9.6×10^{16}	1.9×10^{12}
- volatile portion ≤ 10 %		1.0×10^{13}		1.9×10^{13}		1.9×10^{13}		1.9×10^{13}
- volatile portion ≤ 1 %		1.0×10^{14}		1.9×10^{14}		1.9×10^{14}		1.9×10^{14}
I 129								
- unspecified		2.1×10^{10}		2.1×10^{12}		2.1×10^{13}		2.1×10^{14}
- on silver-containing filters from off gas purification in reprocessing plants		2.1×10^{12}						
Ra 226								
- non-immobilized		1.0×10^{10}		5.2×10^{13}		5.2×10^{14}		5.2×10^{15}
- immobilized		1.6×10^{11}						
further β/γ-emitters		d) 4.2×10^{18} e) 4.2×10^{16}		d) 4.2×10^{20} e) 4.2×10^{18}		d) 4.2×10^{20} e) 4.2×10^{18}		d) 4.2×10^{20} e) 4.2×10^{18}
further α-emitter		2.1×10^{17}		2.1×10^{19}		2.1×10^{19}		2.1×10^{19}

* Concentration of tritium (as HTO) in water, or residual moisture of waste form, ≤ 1.1×10^{13} Bq/m³. Total activity in waste form without tritium activity: a) < 10^{10} Bq; b) 10^{10} Bq to < 10^{12} Bq; c) ≥ 10^{12} Bq. Percentage (mass) of water, or residual moisture in waste form: d) < 1 %; e) ≥ 1 %.

Table 6: Activity limiting values for radionuclides and radionuclide groups disposable per year in the Konrad repository, resulting from the safety assessment of normal operation (in Bq/a)

3.2 INCIDENTS

As is common practice in nuclear enginering, assumed incidents which may occur due to events internal to the plant or influences from the outside must be investigated for a repository within the scope of precautions against damage. Compared with the analyses for nuclear power plants or reprocessing plants, the incident analyses can be carried out by comparatively simple means, as radioactive wastes are only accepted in solid or solidified form and no active components with sophisticated control and supervision systems are therefore required for the inherent activity. Within the scope of an analysis of events, the sequence of operations in the plant is analyzed, from the delivery of the waste packages up to their underground emplacement, with a view to a potential release of radioactive substances, not provided in normal operation, as a result of mechanical and/or thermal influences on the waste packages. On the basis of a classification of individual events in representative groups and an evaluation with regard to the precautions to be taken and the kind of proof to be furnished of such precautions, the courses of events have been assigned to two classes, following the "incident directives" for pressurized water reactors /8/:

- class 1: Incidents whose radiological effects are limited by the design of the plant and/or the waste packages.
- class 2: Incidents which are avoided by design measures for the plant and/or the waste packages.

For incidents of class 1 it must be proved by radiological calculations that the incident dose limits imposed by the authorities under section 28 para. 3 of the Radiation Protection Ordinance are complied with. In the case of incidents of class 2, the respective incident is avoided by precautions the efficiency of which must be proved.

From among about 50 sequences of events which have been analyzed, the following three incidents have been assigned to class 1:

Surface facility:

- drop of transport units in the transfer hall or in the buffer hall from a height of 3 m onto the ground of the hall.

Subsurface facility:

- drop of a waste package in the storage cabin from a height of 5 m, and
- collision of a loaded transport vehicle with subsequent fire.

On the basis of the extent and the efficiency of the precautions to be taken, it must be decided for each individual case to which class the respective incident must be assigned and if appropriate precautions

can better be taken by design measures inside the plant or on the waste packages. In accordance with the calculation methods stated in chapter 4 of the bases for the calculation of incidents /9/, the radiation exposure in the environment of the plant was determined for the above-mentioned incidents resulting from the design; in a further step, activity limitations were derived for radionuclide mixtures in waste packages, attention being paid to the fact that the values do not reach the incident dose limits imposed by the authorities under section 28 para. 3 Radiation Protection Ordinance. The activity limitation can be expressed by the following summation criterion:

$$S_s(p,k) = V \cdot \sum_i \frac{A(i)}{G_s(i,p,k)} < 1 \qquad (3)$$

where:

$S_s(p,k)$ Summation value (s = index for incident)

$A(i)$ Activity of the radionuclide i/the radionuclide group i in a waste package

$G_s(i,p,k)$ Activity limiting value of the radionuclide i/the radionuclide group i, determined from incident analyses for waste form group p and waste packagings class k.

V Factor which takes into account that a transport unit composed of various waste packages may fall ($V = 1$ to 3).

The summation criterion can be applied in two different ways:

- only the activity limiting values for the so-called key nuclides and for non-specified α- and β/γ-emitters are used, or

- the key nuclide activity limiting values and activity limiting values for another 65 single nuclides and for non-specified α- and β/γ-emitters are used.

The key nuclides comprise the 15 radionuclides which are of increased importance in the case of incidents.

Table 7 states the 15 key nuclides for packagings classes I and II. In class I, a distinction is made between six waste form groups. Further details are given in /10/ to /14/.

Radionuclide/ radionuclide-group	Waste class I					Waste class II
	01 Bitumen and plastic products	02 Solid matter	03 Metallic solid matter	04 Compacted waste	05, 06 Cemented/concreted wastes; concentrates	Waste form groups 01 to 06
Cl 36	7.3×10^{8}	7.3×10^{8}	7.3×10^{8}	7.3×10^{8}	7.3×10^{8}	1.2×10^{11}
I 129	2.7×10^{9}	2.7×10^{9}	2.7×10^{9}	2.7×10^{9}	2.7×10^{9}	4.4×10^{11}
Ra 226	4.9×10^{8}	2.4×10^{10}	6.0×10^{10}	1.6×10^{11}	4.9×10^{11}	8.1×10^{12}
Sr 90	1.9×10^{9}	8.9×10^{10}	2.2×10^{11}	5.8×10^{11}	1.9×10^{12}	3.1×10^{13}
Cm 248	2.6×10^{9}	1.3×10^{11}	3.2×10^{11}	8.1×10^{11}	2.6×10^{12}	4.3×10^{13}
Pb 210	3.1×10^{9}	1.6×10^{11}	4.0×10^{11}	9.6×10^{11}	3.1×10^{12}	5.1×10^{13}
Pa 231	3.3×10^{9}	1.6×10^{11}	4.0×10^{11}	1.0×10^{12}	3.3×10^{12}	5.1×10^{13}
Th 232	4.4×10^{9}	2.2×10^{11}	5.6×10^{11}	1.4×10^{12}	4.4×10^{12}	7.3×10^{13}
Ra 228	4.9×10^{9}	2.4×10^{11}	6.2×10^{11}	1.6×10^{12}	4.9×10^{12}	8.1×10^{13}
Pu 239	6.7×10^{9}	3.3×10^{11}	8.1×10^{11}	2.1×10^{12}	6.7×10^{12}	1.1×10^{14}
Ac 227	9.6×10^{9}	4.7×10^{11}	1.2×10^{12}	2.9×10^{12}	9.6×10^{12}	1.6×10^{14}
Tc 99	1.1×10^{10}	5.6×10^{11}	1.4×10^{12}	3.6×10^{12}	1.1×10^{13}	1.9×10^{14}
Am 242m	1.5×10^{10}	7.4×10^{11}	1.9×10^{12}	4.7×10^{12}	1.5×10^{13}	2.4×10^{14}
Cs 137	4.0×10^{10}	1.9×10^{12}	4.9×10^{12}	1.3×10^{13}	4.0×10^{13}	6.7×10^{14}
Co 60	4.9×10^{10}	2.4×10^{12}	6.2×10^{12}	1.6×10^{13}	4.9×10^{13}	8.1×10^{14}
α-emitters	6.7×10^{9}	3.3×10^{11}	8.1×10^{11}	2.1×10^{12}	6.7×10^{12}	1.1×10^{14}
β/γ-emitters	4.9×10^{10}	2.4×10^{12}	6.2×10^{12}	1.6×10^{13}	4.9×10^{13}	8.1×10^{14}

Table 7: Activity limiting values for key nuclides and non-specified further α- and β/γ -emitters resulting from the incident analysis. Data in Bq per waste package

3.3 THERMAL INFLUENCE UPON THE HOST ROCK

The heat generated during the decay of radionuclides leads to a temperature increase in the storage cabins, 7 m in width and about 6 m in height, which, for the planned Konrad repository, should not on an average exceed 3 K at the face of the emplacement room.

This temperature change of the rock is small compared with the changes caused by the ventilation of the mine workings.

In order to meet this emplacement requirement, limiting heat outputs which are related to the unit length of a emplacement room have been derived for the radionuclides /15, 16/. Under the aspect of a uniform formulation of requirements focussed on activities and waste packages, the limiting heat outputs have been converted into package-specific activity limiting values. The activity limitation for a waste package can be indicated by the following summation criterion:

$$S_W(B) = \sum_i \frac{A(i)}{G_W(i,B)} < 1 \qquad (4)$$

where:

S_W Summation value (w = index for heat)

$A(i)$ Activity of the radionuclide i/the radionuclide group i in a waste package

$G_W(i,B)$ Activity limiting value of the radionuclide i/the radionuclide group i, determined from calculations of heat dissipation for the packaging B. G_W values are given in /5/ and /16/.

In this case, the activity limiting values G_W neither depend on the waste form nor on the packagings class, but only on the geometric dimensions of the waste packages. The dimensions of the packages determine the maximum number of waste packages which can be stacked in the cross section of an emplacement room and thus the activity introduced per unit length of the emplacement room. Analogously to the way of proceeding in the case of incidents, two possibilities exist for the application of the summation criterion. The activity limiting values for up to 30 key nuclides and for non-specified α- and β/γ-emitters are used, or these values are combined with those of further 76 radionuclides.

As a mixed emplacement of waste packages with different heat outputs is permissible in the cross section of a storage cabin, waste packages with summation values above 1 can be emplaced, provided that other waste packages show correspondingly lower summation values. In this case, a volume-weighted averaging of the summation values must be carried out. Emplacement is possible if the summation value thus determined is less than one. A further thinning in axial direction of the cabin allows for up to three container lengths. A volume-weighted averaging of the summation values must be carried out also in this case, and it must be guaranteed that the summation value thus determined is less than one.

3.4 CRITICALITY SAFETY

Waste packages containing residual quantities of substances which can be fissioned by thermal neutrons are also to be disposed of in the planned Konrad repository. When such waste packages are delivered to the repository by public carriers, criticality safety is guaranteed by the observance of the transport regulations /17/. In the repository it must be checked if criticality safety is guaranteed during the handling of these waste packages, their storage in a buffer hall and their disposal in emplacement rooms. It must furthermore be investigated if critical fission product distributions may occur in the postoperational phase of the repository due to water access to the wastes and leaching of the fission products from the waste forms.

For the above-mentioned sequences of operation, criticality safety is ensured by a limitation of the mass per unit volume of fissionable products to 50 g per 100 l waste form. Further requirements with regard to the fission product masses in waste packages (Table 8) result from the assumed intrusion of brines into the wastes in the postoperational phase. They can also be expressed in the form of a summation criterion:

$$S_K(B) = \frac{A_{U\ 235}}{G_{U235}(a,B)} + \frac{A_{Pu239}}{G_{Pu239}(B)} < 1 \qquad (5)$$

where:

$S_K(B)$ Summation value (K = index of criticality)

A_{U-235}/A_{Pu-239} Activity of the radionuclide U-235 or Pu-239 in the waste package

$G_{U-235}(a, B)$ Activity limiting value of the radionuclide U-235, determined from criticality calculations for the degree of enrichment a and the packaging B

$G_{Pu-239}(B)$ Activity limiting value of the radionuclide Pu 239, determined from criticality calculations for the packaging B.

Radio-nuclides	Degree of enrichment	Permissible masses		
		Cylindrical packagings	Containers I, II, VI	Containers III, IV, V
U 235	≤ 5 %	120	425	850
U 235	> 5 %	50	175	350
Pu 239	–	28	110	220

Table 8: Permissible masses of fissionable products, resulting from the analyses for criticality safety

Similar to the case of the thermal influence upon the host rock, waste packages with summation values above one can be emplaced if a "thinning" with waste packages of correspondingly lower summation values is ensured in the cross section of one storage cabin. In this case, a number-weighted averaging must be carried out. Emplacement is possible if the summation value thus determined is less than one.

The waste packages (waste form and packaging) which are to be emplaced in the planned Konrad repository must comply with the above-defined requirements of safety engineering. Proof is furnished on the basis of waste package quality control.

LITERATURE

/ 1/ Neufassung des Gesetzes über die friedliche Verwendung der Kernenergie und den Schutz gegen ihre Gefahren (Atomgesetz, Atomic Energy Act) in its version of 31 October 1976. Federal Law Gazette, 1976, part I, No. 131, p. 3053-3072, Bonn, November 6, 1976.

/ 2/ Verordnung über den Schutz vor Schäden durch ionisierende Strahlen (Strahlenschutzverordnung, Radiation Protection Ordinance) of 13 October 1976, Federal Law Gazette, 1976, part I, No. 125, p. 2905-2995, Bonn, October 30, 1976.

/ 3/ Brennecke, P., Schumacher, J.: Anfall radioaktiver Abfälle in der Bundesrepublik Deutschland - Abfallerhebung für das Jahr 1985. PTB-SE-12. Braunschweig 1986.

/ 4/ Sicherheitskriterien für die Endlagerung radioaktiver Abfälle in einem Bergwerk. Federal Law Gazette 35 (1983), No. 2, p. 45-46.

/ 5/ Brennecke P., Warnecke, E.: Anforderungen an endzulagernde radioaktive Abfälle (Vorläufige Endlagerungsbedingungen, Stand November 1986) - Schachtanlage Konrad - PTB informiert: Fachbeiträge zur Sicherstellung und zur Endlagerung radioaktiver Abfälle, 2/87, April 1987.

/ 6/ Lange, F., Müller, W.: Formalism and Models for the Derivation of Acceptance Requirements on Radioactive Waste from Radiation Protection Considerations for the Konrad Waste Repository. In: Merz, E., Odoj, R., Warnecke, E. (editors), Proceedings of the International Seminar on Radioactive Waste Products - Suitability for Final Disposal. Jülich 10.-13.06.85, Berichte der Kernforschungsanlage Jülich, Jül-Conf-54 (1985), p. 622-631.

/ 7/ Ehrlich, D.: Ableitung von Anforderungen an Abfallgebinde aus der Sicherheitsanalyse des bestimmungsgemäßen Betriebes - Aktivitätsbegrenzungen -, in: Physikalisch-Technische Bundesanstalt (editor), Anforderungen an radioaktive Abfälle für das Endlager Konrad und Produktkontrolle. Papers read on the occasion of the 62nd PTB Seminar on 29.10.1985 at the PTB, PTB-SE-8, Braunschweig 1986, p. 102-116.

/ 8/ Leitlinien zur Beurteilung der Auslegung von Kernkraftwerken mit Druckwasserreaktoren gegen Störfälle im Sinne des § 28 Abs. 3 StrlSchV - Störfall-Leitlinien - Federal Law Gazette 35 (1983), No. 245 a, p. 3-9.

/ 9/ Störfallberechnungsgrundlagen für die Leitlinien des BMI zur Beurteilung der Auslegung von Kernkraftwerken mit DWR gemäß § 28 Abs. 3 StrlSchV. Federal Law Gazette 35 (1983), No. 245 a, p. 11-24.

/10/ Illi, H.: Sicherheitsaspekte bei der Endlagerung, in: Kerntechnische Gesellschaft e. V. (editor): Berichtsband zum Fortbildungsseminar Neuere Entwicklungen zur Sicherheit von Kernkraftwerken und Anlagen des Kernbrennstoffkreislaufs, Düsseldorf, 24.-25.09.1985, Düsseldorf 1985, p. 306-344.

/11/ Berg, H. P., Illi, H.: Störfallanalysen für das Endlager Konrad, PTB-Mitteilungen 95 (1985) No. 6, p. 407-413.

/12/ Lange, F., Wurtinger, W.: Vorgehen bei den Störfallanalysen für die Schachtanlage Konrad, in: Gesellschaft für Reaktorsicherheit mbH (editor), 8. GRS-Fachgespräch, Köln, 12.-13.11.1984, Bericht GRS-58, Köln 1985, p. 52-59.

/13/ Gründler, D.: Accidental Release of Radioactivity from Waste Packages in a Final Waste Repository, in: Post, R. G. (editor), Proceedings of the Symposium on Waste Management '85, Tucson, 24.-28.03.1985, vol. 3, (1985) p. 183-188.

/14/ Wurtinger, W.; Gründler, D., Mielke, H., Sütterlin, L.: Accidental Release from Radioactive Waste Packages, in: Transportation Research Board/National Research Council (editors): Proceedings of Recent Advances in Hazardous Materials Transportation Research - An International Exchange, Lake Buena Vista, 10.-13.11.1985, TRB State-of-the-Art Report 3, Washington 1986, p. 3-8.

/15/ Piefke, F.: Berechnungen zur thermischen Einwirkung von schwachwärmeentwickelnden radioaktiven Abfällen auf das Wirtsgestein in der Schachtanlage Konrad. PTB-SE-10. Braunschweig 1986.

/16/ Piefke, F.: Anforderungen an Abfallgebinde aufgrund der thermischen Beeinflussung des Wirtsgesteins der Schachtanlage Konrad. PTB-SE-14. Braunschweig 1986.

/17/ Verordnung über die Beförderung gefährlicher Güter auf der Straße (Gefahrgutverordnung Straße - GGVS), Anlagen A und B. Annex to Federal Law Gazette part 1, No. 55 of 31.08.1979.

The planned GORLEBEN repository – present status, new data provided by the utilities on wastes produced, and future directions –

E. Warnecke, K. Ermisch

SUMMARY

It is planned to construct and operate a repository for all types of solid or solidified radioactive wastes in the Gorleben salt dome, Federal Republic of Germany.

On the basis of model assumptions and partially site-specific data, preliminary safety assessments were undertaken. The site has been investigated from above-ground, resulting in its presumed suitability for the disposal of radioacitve wastes. The shaft sinking therefore begun, followed by an underground investigation. These results are necessary for a final judgement on the suitability of the salt dome for waste disposal.

As intermediate step in the elaboration of the application documents a new data basis for the radioactive wastes to be disposed of was drafted on the basis of experience gained with the Konrad mine. These include:

- Packagings: Weight of waste packages up to 40 Mg; 8 packagings for heat generating wastes have been standardized;

- Wastes: About 20.000 m^3 per year of non-heat generating radioactive wastes and heat generating wastes including about 115 m^3 of glass blocks, 1400 m^3 of hulls, ends etc., 116 m^3 of HTR spent fuel and spent LWR fuel in Pollux packagings or ingot moulds are assumed;

- Disposal techniques: Non-heat generating radioactive wastes are planned to be stacked in drifts; the heat generating radioactive wastes will be disposed of in boreholes of a depth of up to 600 m.

Dr. E. Warnecke, Physikalisch-Technische Bundesanstalt, Bundesallee 100, Postfach 33 45, D-3300 Braunschweig, Federal Republic of Germany; K. Ermisch, Preußenelektra AG, D-3000 Hannover, Federal Republic of Germany

1 INTRODUCTION

In the Federal Republic of Germany, the Physikalisch-Technische Bundesanstalt (PTB) is responsible for the construction and operation of the repositories for radioactive wastes /1/ and is supported by the Bundesanstalt für Geowissenschaften und Rohstoffe (BGR). The PTB is directly supervised by Federal authorities and may make use of the services of third parties in order to construct or operate these facilities. For this task, the Deutsche Gesellschaft zum Bau und Betrieb von Endlagern für Abfallstoffe mbH (DBE) was founded in 1979. The organization of the disposal of radioactive wastes is illustrated in Fig. 1.

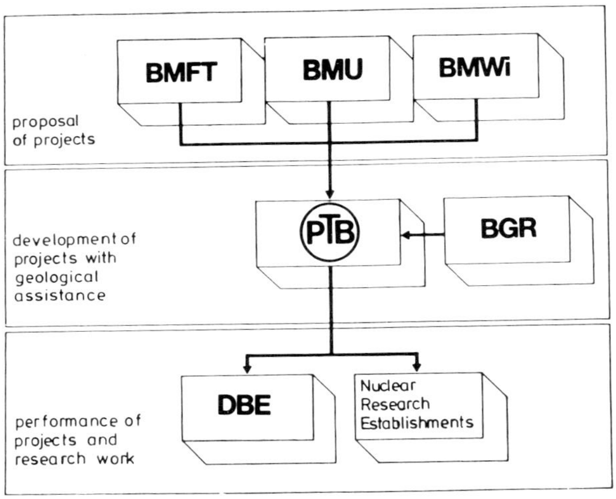

Fig. 1: Chain of responsibility

The proper conditioning of radioactive waste for disposal is the task of the waste producers. They also must provide an adequate data basis of the wastes produced in order to plan the necessary disposal capacity and the layout of the facilities for the types and amounts of waste produced. The existing results and the direction of future efforts based on the figures of the utilities on wastes produced will be discussed.

Two repositories are planned (Fig. 2):

- Gorleben (salt dome; for all types of radioactive wastes)

- Konrad (abandoned iron ore mine; for radioactive wastes with negligible thermal influence upon the host rock).

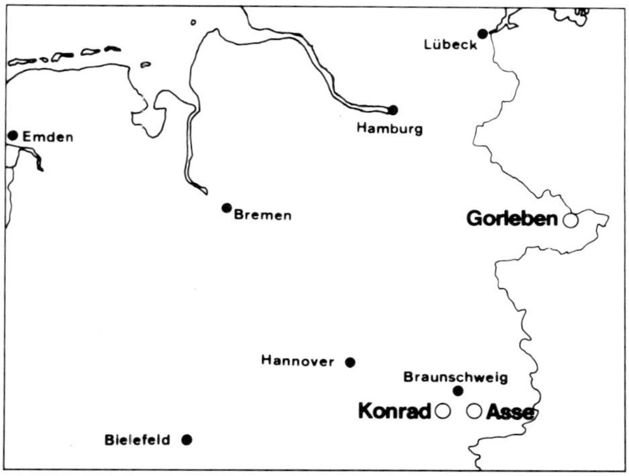

Fig. 2: Locations of the site

Additionally, it is planned to inject tritium-containing water from the reprocessing of spent fuel into deep geological formations, especially porous sediments. A preselection of possible sites in Bavaria is under way.

From 1967 to 1978, the former Asse salt mine was used as a pilot repository within the scope of a licence limited in time. It now serves as the R+D facility and aims especially at determining the thermal and thermomechanical effects of the heat generating wastes. This information is required for the construction of the Gorleben repository.

2 SCOPE

It is the purpose and the aim of the current work to construct and operate a repository for all types of solid or solidified wastes in the Gorleben salt dome in about the year 2000.

The principal procedures and the elaboration of the application documents must be in compliance with the "Safety Criteria for the Disposal of Radioactive Wastes in a Mine" /2/. These criteria stipulate in particular,

- that the required safety of a repository must be demonstrated by a site-specific safety assessment which includes the overall geological situation, the technical concept of the disposal mine and the waste packages;

- the objectives of the operation of a repository which are prescribed by the German Atomic Energy Act /1/;

- the radiation protection targets for repositories which are quantified in the German Radiation Protection Ordinance /3/.

The required site-specific proof of the repository's safety necessitates a thorough site investigation, a detailed layout of the mine and an accurate planning of the types and amounts of wastes and its emplacement locations.

Such an ambitious work can only be carried out in several steps beginning with a drafting of concepts on model data and ending with a final plan of the repository after a completed site investigation and safety assessment. An intermediate step in the elaboration of the application documents is at present being initiated for the Gorleben facility. Its purpose it to

- take into account the actual developments in nuclear industry and waste management;

- incorporate the present site-specific knowledge into the repository plans;

- assess the specifications of COGEMA and BNFL for wastes to be returned;

- develop a yardstick to rate the results of the underground exploration of the salt dome.

Knowledge of the repository to date and the status of the work will be summarized. Trends and directions for the intermediate plannings will also be outlined.

3 THE PLANNED GORLEBEN REPOSITORY

3.1 EXISTING PLANNING AND DATA

All existing assessments of the Gorleben site are based on model assumptions of the geology, the layout of the repository and the radioactive wastes. They were performed at the end of the seventies. Only the calculations of the long-term safety include parts of the hydrogeological investigations, therefore only weak points could be identified at this preliminary stage. All the assessments will have to be updated after the underground exploration. The disposal of spent fuel will also have to be allowed for in future planning.

3.1.1 SITE INVESTIGATION

The Gorleben site is under investigation to check the suitability of the salt dome for the disposal of all types of radioactive wastes, especially heat-generating wastes. A comprehensive site investigation programme has been elaborated in cooperation with BGR /4/.

The investigations above ground are almost completed /5/ and the results were published in May 1983 in a summarizing interim report /6/.

These investigations were carried out in an area of roughly 300 km^2 and included:

- 129 hydrogeological exploratory drillings,
- 275 observation drillings, which have been expanded for use as ground-water measuring locations,
- 3 wells for pumping tests,
- 11 core drillings,
- more than 40 drillings sunk down to 80 m into the salt,
- 4 salt-dome exploratory drillings of about 2000 m into the flanks of the salt dome,

- an intensive seismic measuring programme, and
- investiations into the retardation of radionuclides on site specific samples.

Fig. 3 shows the exploration area and the locations of the hydrogeological investigations, the 4 salt dome exploratory drillings and 2 shaft pre-borings.

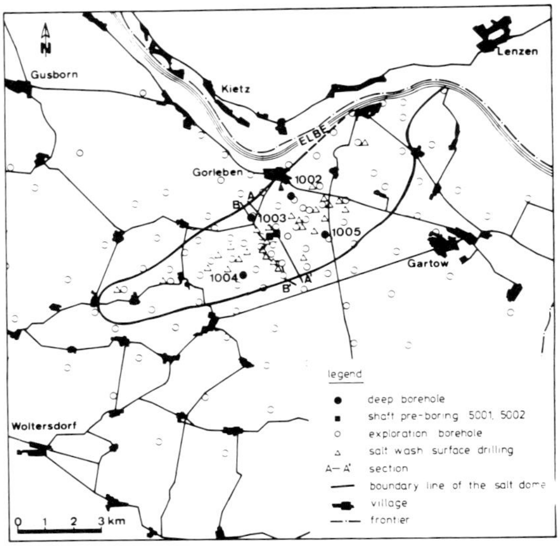

Fig. 3: Locations of exploratory drill holes

The resulting site-specific data include special information on

- the hydrogeological situation in the overlying and surrounding strata of the salt dome,
- the sorption data of radionuclides on Gorleben samples (sediment and waters) /7,8,9/,
- the geological situation in the caprock area,
- the existence and sequence of the various strata within the salt dome, and
- the locations of the shafts.

Figures 4 and 5 are sectional views of the overlying strata and the salt dome as examples of the results of the site investigation.

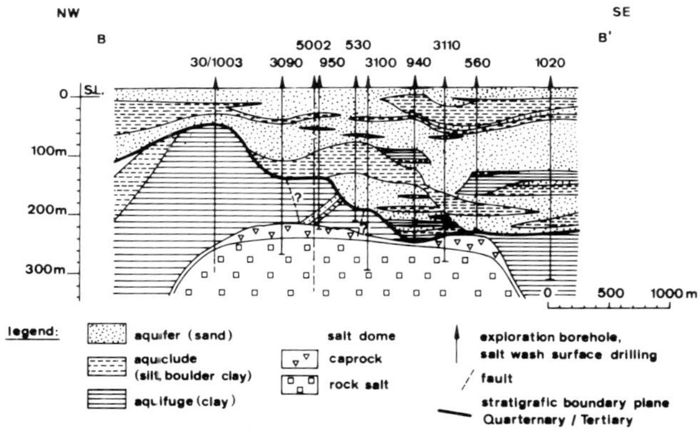

Fig. 4: Geological section of the overlying strata in the channel area

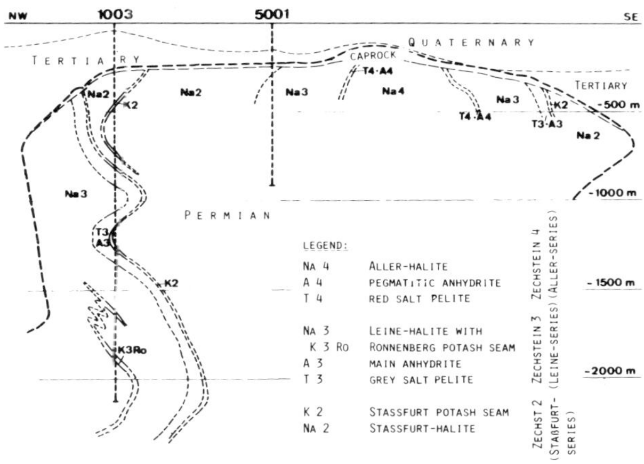

Fig. 5: Section of salt dome

It was concluded from these results that the salt dome's suitability for waste disposal was confirmed and that those data which can only be obtained by an undergrond exploration are sill missing. For this purpose, two shafts must be sunk and then the salt dome is to be explored by drillings from galleries in the exploratory mine into the surrounding rock /10/.

In July 1983, the Federal Government approved the speedy commencement of the underground exploration.

The two shafts are being sunk by the Arbeitsgemeinschaft Schächte Gorleben, ASG (Gorleben shafts pool) making use of the "freezing technique". The work has been in progress since 1984. The preparation of the "freezing boreholes" began in May 1984 (Fig. 6). After the construction of a frost jacket, the work on the sinking of the first shaft began in March 1986.

Fig. 6: Sinking of shafts by "freezing technique" (Gorleben)

The shaft-sinking activities were interrupted at shaft 1 due to an accident which occurred in May 1987 as the shaft was at a depth of about 239 m. This was obviously a consequence of convergence within a clay layer. A preliminary liner which was installed to prevent convergence fell, killing one miner and injuring 6.

The convergence was halted by refilling about 14 m of the shaft with lean concrete. A concept for the stabilization of the clay layer is now being elaborated, and it is intended to continue shaft sinking on the basis of this concept. In principle, the present situation is known in conventional mining and shaft sinking /23/ and does not influence the basic suitability of the salt dome for the disposal of radioactive wastes.

The sinking of the pre-shaft of the second shaft has been halted at a depth of about 39 m. It is anticipated that shaft-sinking activities will be continued when the discussions on the clay layers are concluded. The delay in shaft sinking due to the accident in shaft 1 will be at least about 1 year.

Detailed exploration of the salt dome will follow by means of exploratory drifts and drillings at a depth of about 840 m. About 120 km of core drilling over an area of about 10 km by 1 km /10/ is planned.

3.1.2 MODEL PLAN FOR THE REPOSITORY

The existing plans were based on data from 1976 on a reprocessing plant with a capacity of 1400 t/year and the corresponding capacity of nuclear power plants. The reprocessing wastes from an electricity generation of 2500 GWa was assumed for disposal as well as 40 - 50 % of the wastes from the nuclear fuel cycle. the cumulative inventory of β/γ-emitters and α-emitters in the repository after 50 years of operation has been estimated on this basis to be in the order of magnitude of 10^{21} Bq and 10^{19} Bq, respectively /6,11/. The annual data for the amounts of waste, the required host rock volume and the activities are compiled in Table 1, /5,11,24/.

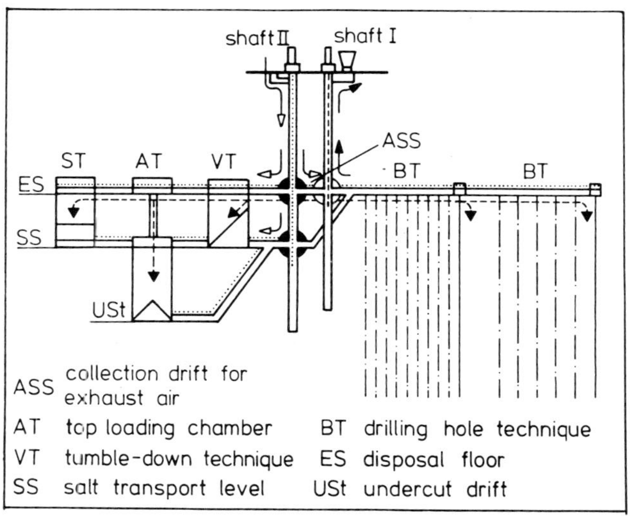

Fig. 7: Scope of the planned mine shown schematically

It was planned to dispose of the heat-generating wastes in boreholes (300 - 600 m deep) and the other wastes in mined rooms (Fig. 7).

	Waste volume produced per annum		Repository volume required per annum		Activity inventory of waste produced per annum	
	m^3	%	m^3	%	Bq	%
Reprocessing	$1{,}25 \cdot 10^4$	39,0	$6{,}45 \cdot 10^6$	97,6	$21{,}20 \cdot 10^{18}$	99,99
HLW portion	$1{,}5 \cdot 10^2$	0,5	$5{,}28 \cdot 10^6$	79,9	$17{,}02 \cdot 10^{18}$	80,4
Nuclear power	$1{,}62 \cdot 10^4$	50,4	$1{,}28 \cdot 10^5$	1,95	$21{,}16 \cdot 10^{14}$	
Research centres	$2{,}21 \cdot 10^3$	6,8	$1{,}64 \cdot 10^4$	0,25	$8{,}14 \cdot 10^{13}$	0,01
Industry	$5{,}63 \cdot 10^2$	1,7	$5{,}36 \cdot 10^3$	0,1		
Collecting depots of the federal states	$7{,}31 \cdot 10^2$	2,3	$6{,}82 \cdot 10^3$	0,1	$34{,}04 \cdot 10^{12}$	

Table 1: Prospective quantity of radioacitve waste for long-term repository planning

These planning data are currently being revised and will be updated to include new developments in the data basis for radioacitve wastes and these results of the site investigation (see chapter 3.2).

3.1.3 SAFETY ASSESSMENTS

a) Normal Operation

In the normal operation of the repository, significant releases from radioactive wastes could result from gaseous radionuclides and aerosols. In a preliminary assessment /12/ it was found that the release of H-3 from zircaloy fuel element hulls fixed in concrete could possibly exceed the limits for the H-3 concentration in the air. Ranges of possibly tolerable average release rates of volatile radionuclides from waste packages have been derived from model calculations (Table 2), /13,24/.

	1	2	3	4	5	6	7
Nuclide	H-3	C-14	Kr-85	Ru-106	J-129	Rn-220	Rn-222
Non-bore-hole technique	$3 \cdot 10^{-3\pm1}$	$2 \cdot 10^{-3\pm1}$	$1 \cdot 10^{-3\pm1}$	$5 \cdot 10^{-6\pm1}$	$1 \cdot 10^{-6\pm1}$	$1 \cdot 10^{-5\pm1}$	$5 \cdot 10^{-5\pm1}$
Bore-hole technique	$5 \cdot 10^{-2}{}^{+0}_{-1}$	$4 \cdot 10^{-2}{}^{+0}_{-1}$	$3 \cdot 10^{-2}{}^{+0}_{-1}$	$6 \cdot 10^{-5}{}^{+0}_{-1}$	$2 \cdot 10^{-5}{}^{+0}_{-1}$	$2 \cdot 10^{-4}{}^{+0}_{-1}$	$1 \cdot 10^{-3}{}^{+0}_{-1}$

Table 2: Preliminary maximum permissible mean annual release rates per waste package for volatile individual radionuclides (release rates in Ci/a waste package)

The exposure to direct radiation can be adjusted to the required limits, e. g. by providing adequate shielding.

b) Incidents

The analysis of possible incidents showed that mechanical or thermal impacts, or a combination of both, could be representative /12,14/. The boundary conditions for these impacts will be evaluated within the scope of the anticipated updating of the preliminary assessments.

c) Thermal aspects

One of the main points in the disposal of heat generating wastes is the influence of the dissipated heat on the waste packages, the disposal area and the host rock.

The layout of the repository is based on a maximum temperature of 200° C at the surface of the glass blocks of the vitrified fission product concentrate and 100° C for cemented wastes /12,14/. This limitation resulted from the chemical resistance of glass in brines and the cement properties, respectively. It has been planned to dispose of these wastes in deep boreholes (300 m). The limitation of the initial heat per waste package is dependent on , e. g., the age of the waste, the distance and length of the boreholes, and the emplacement within the boreholes (longitudinal heat dilution) /14,15,16/.

It is very important that the highest temperatures occur only locally in the center of a borehole field and for a limited period of about 100 years (between 50 and 150 years after disposal). The temperature outside the disposal field remains below 100° C. The natural temperature distribution in the host rock is almost reached 5000 years after waste disposal.

The effects of this temperature distribution have been simulated with two independent numerical computer programmes /17,18/. The thermal expansion of the salt induces a calculated uplift of 1,2 m above the repository 450 years after disposal (Fig. 8), /17/. From these investigations no arguments could be found against this layout of the model repository. A concept for the final layout can only be made after the underground investigation when the questions relevant to geochemical, physical, chemical and petrophysical aspects can be answered.

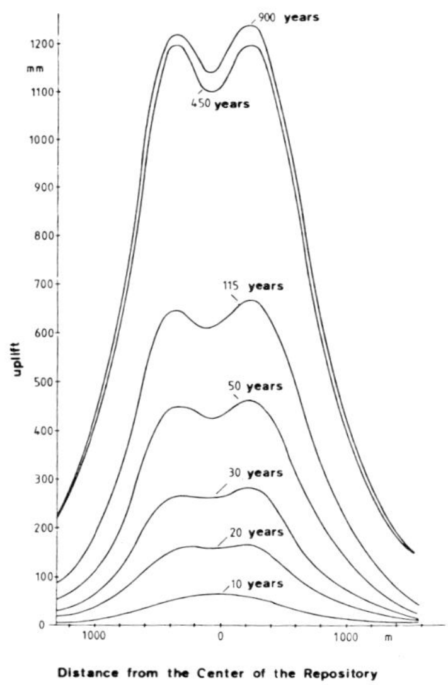

Fig. 8: Uplift above the repository /17/

d) Long-Term Safety Assessment

The intrusion of the salt into the younger, overlying strata took place between the geological periods of Malm and Lower Cretaceous. As a result of diapirism, the salt layers are folded.

Detailed knowledge of the interior of the salt dome will only be available after its underground investigation. Possible incidents are therefore conservatively considered in theory for the long-term safety assessment. It is assumed that thermomechanical effects caused by heat-generating radioactive wastes can produce new pathways for waters in the anhydrite horizons in the post-operational phase of the repository (especially in the Hauptanhydrit). As a result of this, water from the ground-water-bearing overlying strata could intrude into the repository area. Contaminated brines could be released back into the overlying strata in the same way, due to the convergence of the rock salt /19/.

In the overlying strata of tertiary, especially quaternary geological age, the sorption of radionuclides will then influence the migration of the radionuclides in the flow of ground-water. Preliminary results from an R+D study on the consequences of such a scenario have been published /20/. The intrusion of waters into the repository area and the releases of contaminated waters are shown in Fig. 9. The calculated doses from these releases are shown in Fig. 10 /20/.

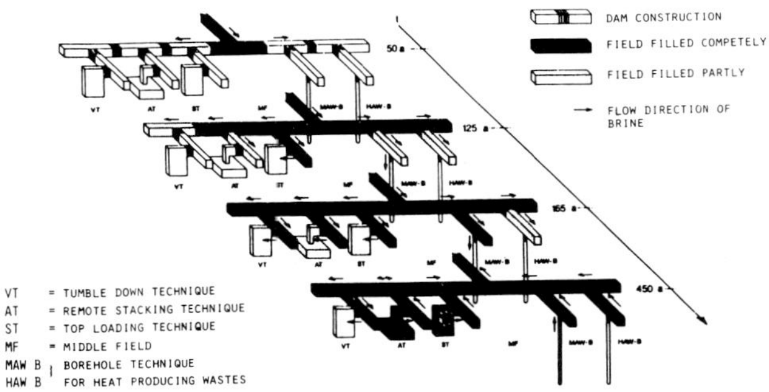

Fig. 9: Time schedule of brine flow /20/.

Some of the main results of this investigation are,

- that the intruding waters do not reach the disposal area for the vitrified fission product concentrates because of the thermal layout,

- that the calculated dose rates in the environment are dominated by Tc and Np,

- that the highest dose rates occur about 10 000 years after disposal, and

- that the safety criteria requirements /9/ are fulfilled.

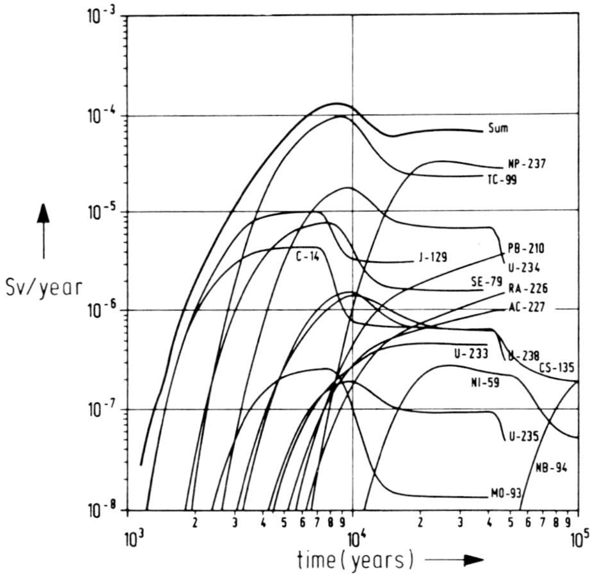

Fig. 10: Contribution of some radionuclides to potential and total radiation exposure as a function of time /20/

Recent results of the investigation into the sorption behaviour of radionuclides in Gorleben samples /8,21,22/ have shown that the sorption data for Np and Tc are considerably higher than assumed in /20/. The dose rates of these radionuclides will therefore be of less importance.

3.2 UPDATING OF THE PLANS FOR THE GORLEBEN REPOSITORY

3.2.1 RADIOACTIVE WASTES

a) Nuclear Installations

In the present planning stage, it is assumed that about 30 light-water reactors and a high-temperature reactor will be in operation in an average year (reference year) of operation of the planned repository. It is envisaged that this nuclear capacity be covering for the repository's period of operation.

It is assumed that the spent fuel from these reactors will be processed via an adequate reprocessing capacity and a conditioning plant for spent fuel as an additional step to reprocessing. The reprocessing of spent fuel in foreign countries is taken into account.

A corresponding fuel fabrication capacity is considered, as well as research establishments and the use of radionuclides in medicine and industry.

b) Packagings, Types and Volumes of Waste Package

A standardization of the packagings for waste disposal in Gorleben has been carried out jointly by the PTB and representatives of the various waste producers. On the basis of the existing standardization of packaging for the planned Konrad repository /25/, the following modifications were made (table 3):

- inclusion of container Type VII and VIII in the list of packagings for wastes with negligible heat generation;

- limitation of the mass of non-heat-generating waste packages to 40 Mg;

- addition of 8 packagings for heat generating wastes, where 3 packagings are limited to its use either in the WAK, the COGEMA or BNFL.

The types of <u>non-heat-generating waste packages</u> are in accordance with the basis data for the Konrad project /11,26-30/. This means that

No.	Designation	Length/ Diameter (mm)	Width (mm)	Height (mm)	Gross Volume (m^3)	Mass (Mg)
01.	Cylindrical concrete packaging Type I	ø 1060	—	1370 1)	1,2	3...4
02.	Cylindrical concrete packaging Type II	ø 1060	—	1510 2)	1,3	3...4
03.	Cylindrical concrete packaging Type III	ø 1400	—	2000	3,1	10..13
04.	Cylindrical cast iron packaging Type I	ø 900	—	1150	0,7	3...6
05.	Cylindrical cast iron packaging Type II	ø 1060	—	1500 3)	1,3	7..12
06.	Cylindrical cast iron packaging Type III	ø 1000	—	1240	1,0	3...6
07.	Container Type I	1600	1700	1450 4)	3,9	≤ 40
08.	Container Type II	1600	1700	1700	4,6	≤ 40
09.	Container Type III	3000	1700	1700	8,7	≤ 40
10.	Container Type IV	3000	1700	1450 4)	7,4	≤ 40
11.	Container Type V	3200	2000	1700	10,9	≤ 40
12.	Container Type VI	1600	2000	1700	5,4	≤ 40
13.	Container Type VII	2000	2000	1600	6,4	≤ 40
14.	Container Type VIII	2800	2000	1600	9,0	≤ 40
15.	Ingot Mould Type I	430	—	1335 5)	0,19	≈ 0,5
16.	Ingot Mould Type II	430	—	2600 5)	0,38	≈ 2
17.	BT-packaging Type I 6)	624	—	927	0,28	≈ 1,25
18.	BT-packaging Type II 7)	775	—	1150	0,54	≈ 2
19.	BT-packaging Type III 8)	800	—	1185	0,60	≈ 2
20.	BT-packaging Type IV 9)	1136	—	1707	1,73	≈ 4
21.	SF-packaging Type I	1500	—	5500	9,7	≈ 64
22.	SF-packaging Type II	1600	—	5000	10,0	≈ 64

1) Height: 1370 mm + 90 mm lifting lug = 1460 mm
2) Height: 1510 mm + 90 mm lifting lug = 1600 mm
3) Height: 1370 mm for KfK type
4) Stacking height: 1400 mm for KfK type
5) Stacking height is 80 mm less
6) WAK-packaging with mushroom lid
7) WAW-packaging with mushroom lid
8) BNFL-packaging
9) COGEMA-packaging
Konrad packagings are limited to a maximum weight of 20 Mg.

Table 3: Survey of Standardized Cylindrical Packagings and Containers for Radioactive Wastes

experience already gained in waste minimization and volume reduction is being introduced into the Gorleben planning. The volume of waste packages with negligible heat generation is assumed to be about 20.000 m^3 for the reference year.

These figures include wastes from the <u>decommissioning of nuclear power plants.</u> It is assumed that on an average, every three years a nuclear power plant is decommissioned.

The <u>heat-generating wastes from reprocesing</u> of spent fuel result in glass blocks from the vitrification of the liquid high level waste concentrate packaged in the ingot mould and hulls, ends etc. in borehole technique packagings (BT packagings). The volumes of these waste packages are assumed to be about 115 m^3 (glass blocks) and 1400 m^3 (hulls, ends etc.) for the reference year.

The <u>disposal of spent fuel</u> is introduced into the Gorleben data basis for waste packages. The spent fuel comprises the graphitic pebbles in BT packagings from the high temperature gas cooled reactor line (HTR) and light-water reactor fuel (LWR). The LWR fuel consists of high burn-up uranium or mixed oxide fuel (MOX) and fuel made from reprocessed uranium.

The LWR fuel elements can be conditioned either by packing the fuel pins and the structural material into thick-walled spent fuel packagings (S.F. packaging, e. g. Pollux packagings), or by packing the cut fuel pins into an ingot mould. The following waste package volumes are assumed for the reference year:

- HTR fuel: about 116 m^3
- LWR fuel: about 225 m^3
  plus about 100 m^3 of waste in containers
  or: about 50 - 70 m^3 in ingot moulds
  plus about 190 m^3 of waste in containers.

3.2.2 DELIVERY, TRANSSHIPMENT AND BUFFERING OF WASTE PACKAGES

The delivery of waste packages is planned either with 100 % use of rail or with equal use of rail and road.

A facility must be planned for the transshipment of waste packages from collective packagings into single casks which are used within the repository.

A buffering capacity of about 2 weeks is necessary to decouple the delivery and the disposal of waste packages.

3.2.3 LAYOUT OF THE REPOSITORY AND DISPOSAL TECHNIQUES

It is assumed that radioactive waste packages from an electricity generation of 2500 GWa can be disposed of in Gorleben. The capacity of the planned repository and the period of operation are dependent on the underground investigation of the salt dome, and attempts to minimize the radioactive wastes. They can therefore only be determined in the nineties.

Although there is at present a very limited knowledge of the internal structure of the salt dome, it seems sufficient to plan deep disposal areas in narrow bands parallel to the flanks.

a) Non-Heat-Generating Wastes

Waste packages with negligible heat generation are planned to be disposed of by stacking techniques in drifts. The introduction of this concept seems to be advantageous because

- there is harmonization with the disposal techniques to be used in Konrad /30/
- the waste volumes have been reduced. The construction of large disposal rooms for various disposal techniques would result in an undesirable long-term opening up of these rooms.

If necessary, it is possible to operate the galleries of the repository at different levels.

b) Heat-Generating Wastes

It is planned to dispose of the heat generating wastes in boreholes of a depth of up to 600 m. Only the spent fuel in thick-walled packagings is intended to be emplaced in drifts, preferably above the boreholes of the vitrified reprocessing wastes.

The thermal layout of the repository area for the heat-generating wastes will decide the distances which are necessary between the boreholes and the waste packages to observe the maximum permissible temperatures.

4 FINAL REMARKS

After preparation of the application documents for the planned Konrad repository, it is intended to update the plans for the Gorleben repository on the basis of present site-specific knowledge. This is a further step towards the formulation of waste acceptance requirements for a Gorleben repository and the preparation of application documents.

The work necessary to realize repositories is in progress in the Federal Republic of Germany and although shaft sinking in Gorleben is at present interrupted, there has been an intensive scientific, technical and engineering programme under way for many years, resulting in the Federal Republic of Germany's leading role in the disposal of radioactive wastes.

LITERATURE

/1/ Neufassung des Gesetzes über die friedliche Verwendung der Kernenergie und den Schutz gegen ihre Gefahren (Atomgesetz) vom 31. Oktober 1976, Bundesgesetzblatt, Jahrg. 1976, Teil I, Nr. 131, Bonn,. 6. November 1976, 3053.

/2/ Sicherheitskriterien für die Endlagerung radioaktiver Abfälle in einem Bergwerk, Bundesanzeiger 35 (1983), No. 2, 45.

/3/ Verordnung über den Schutz vor Schäden durch ionisierende Strahlen (Strahlenschutzverordnung - StrlSchV) vom 13. Oktober 1976, Bundesgesetzblatt, Jahrg. 1976, Teil I, No. 125, Bonn, 30. Oktober 1976, 2905.

/4/ H. Röthemeyer, Site Investigations and Conceptual Design for the Repository in the Nuclear "Entsorgungszentrum" of the Federal Republic of Germany, Int. Symp. on Underground Disposal of Radioactive Wastes, Helsinki, July 2-5, 1979, Vol. I, IAEA, Vienna, 1980, 297.

/5/ H. Röthemeyer, A. V. Joshi, B. Verkerk, Establishing Deep Geological Repositories for Radioactive Wastes with Particular Reference to Disposal in Salt Formations, Int. Conf. on Radioactive Waste Management, May 16 - 30, 1983, Vol. 3, IAEA, Vienna, 1984, 49.

/6/ Zusammenfassender Zwischenbericht über bisherige Ergebnisse der Standortuntersuchung in Gorleben, Physikalisch-Technische Bundesanstalt, Braunschweig, Mai 1983.

/7/ E. Warnecke, A. Hollmann, G. Stier-Friedland; Migration of Radio nuclides, Experiments within the Site Investigation Program at Gorleben, in: McVay, G. L. (ed.): Scientific Basis for Nuclear Waste Management, North-Holland, New York/Amsterdam/Oxford, 7 (1984), 41.

/8/ E. Warnecke, G. Tittel, P. Brennecke, G. Stier-Friedland, A. Hollmann, Experimental Investigations of Possibly Occurring Radionuclide Releases from the Planned Repositories in the Gorleben Salt Dome and the Konrad Iron Ore Mine for the Long-Term Safety Assessment, Proc. Int. Symp. on the Siting, Design and Construction of Underground Repositories for Radioactive Wastes, Hannover, March 3 - 7, 1986, IAEA Vienna 1986, 401.

/9/ E. Warnecke, W. Hild, German Experience in the Field of Radionuclide Migration in the Geosphere, Int. Conf. on Chemistry and Migration Behaviour of Actinides and Fission Products in the Geosphere, Munich, September 14 - 18, 1987, Radioactive Waste Management and the Nuclear Fuel Cycle, Vol. 10 (1 - 2), 1988, 115.

/10/ G. Grübler, Programme for Investigating the Suitability of the Gorleben Salt Dome as a Repository for Radioactive Wastes, Proc. Int. Symp. on the Siting, Design and Construction of Underground Repositories for Radioactive Wastes, Hannover, March 3 -7, 1986, IAEA, Vienna, 1986, 363.

/11/ P. Brennecke, E. Warnecke, Disposal of Radioactive Wastes from Nuclear Power Plants in the Federal Republic of Germany, Proc. IAEA Sem. on Management of Radioactive Waste from Nuclear Power Plants, Karlsruhe, October 05 09, 1981, IAEA-TECDOC-276, IAEA, Vienna, 1983, 213.

/12/ E. Warnecke, H. Illi, Evaluation of Product Specifications with a Safety Analysis for a Disposal Mine, in: J. G. Moore (ed.), Scientific Basis for Nuclear Waste Management, Vol. 3, Plenum Press, New York/London, 1981, 19.

/13/ P. Brennecke, D. Ehrlich, Radioaktive Abfälle mit flüchtigen Nukliden aus der Sicht der Endlagerung in Bergwerken, in: Ergebnisse eines Strategiegespräches über Forschungs- und Entwicklungsarbeiten zur Entsorgung von radioaktiven Sonderabfällen, Bonn 24.-25. August, 1983; Bundesminister für Forschung und Technologie, Bonn.

/14/ E. Warnecke, H. Illi, D. Ehrlich, Requirements for the Disposal of High-Level Radioactive Wastes, in: R. Odoj/E. Merz (eds.), Proc. Int. Sem. on Chemistry and Process Engineering for High Level Liquid Waste Solidification, Jülich, June 1-5, 1981, Bericht der Kernforschungsanlage Jülich, Jül-Conf-42 (Vol. 2), Jülich, 1981, 792.

/15/ G. Delisle, Berechnung der raumzeitlichen Entwicklung des Temperaturfeldes um ein Endlager für mittel- und hochaktive Abfälle in einer Salzformation, Z. dt. geol. Ges. 131, 1980, 461.

/16/ F. Piefke, Berechnungen zur Aufheizung einer Salzformation durch ein HAW-Endlager bei Berücksichtigung nicht-wärmeentwickelnder Bohrlochabschnitte, Teil 1, Mathematische Grundlagen und Ergebnisse für nicht-wärmeentwickelnde Bohrlochabschnitte aus Salz; Bericht im Auftrag der PTB, September 1982.

/17/ F. Rischbieter, R. A. Hintz, K. Michelmann, E. Schäpermeier, Geologische Aspekte bei der Schwachstellenbetrachtung für das Bundesendlager für radioaktive Abfälle am Standort Gorleben, Battelle-Institut e.V., Frankfurt am Main, Bericht BF-R-63.861-2, 1980.

/18/ H. Albrecht, U. Hunsche, Gebirgsmechanische Aspekte bei der Endlagerung radioaktiver Abfälle in Salzdiapiren unter besonderer Berücksichtigung des Fließverhaltens von Steinsalz. Fortschr. Miner. 58, 1980, 212.

/19/ Störfälle als Folge des Zuflusses von Wässern oder Salzlösungen in ein Salinar-Bergwerk in steiler Lagerung für die Endlagerung radioaktiver Abfälle, PTB Info-Blatt 4/82, 1982.

/20/ Entwicklung eines sicherheitsanalytischen Instrumentariums für das geologische Endlager für radioaktive Abfälle in einem Salzstock, Zusammenfassender Abschlußbericht, Kapitel 4, Projekt Sicherheitsstudien Entsorgung, Berlin, 1985.

/21/ M. Mühlenweg, K. H. Lieser, Neptunium in the Geosphere, Int. Conf. on Chemistry and Migration of Actinides and Fission Products in the Geosphere, Munich, September 14 - 18, 1987.

/22/ C. Hanke, K. H. Lieser, Technetium in the Geosphere, Int. Conf. on Chemistry and Migration of Actinides and Fission Products in the Geosphere, Munich, September 14 - 18,1987.

/23/ L. Gebhardt, Die Anwendung des Gefrierverfahrens; Festschrift zur 25. Wanderversammlung der Bohr-Ingenieure, Budapest 1911, 114.

/24/ E. Warnecke, The German Approach to the Disposal of Radioactive Wastes, Finnish-German Seminar on Nuclear Waste Management, Otaniemi, September 23 - 25, 1986 (to be published).

/25/ P. Brennecke, E. Warnecke, Anforderungen an endzulagernde radioaktive Abfälle (Vorläufige Endlagerungsbedingungen, Stand November 1986) - Schachtanlage Konrad, Bericht PTB-SE-16, Januar 1987.

/26/ E. Warnecke, P. Brennecke, G. G. Eigenwillig, F. Heil, Quality Control of Radioactive Wastes, in: Proc. Int. Symp. on Conditioning of Radioactive Wastes for Storage and Disposal, Utrecht,: June 21 - 25 1982, IAEA, Vienna, 1983, 391.

/27/ P. Brennecke, D. Ehrlich, H. Illi, E. Viel, E. Warnecke, Radioaktive Abfälle endlagergerecht - Erfassung und Kategorisierung radioaktiver Abfälle. Energiewirtschaftliche Tagesfragen 34, 1984, Nr. 10, 767.

/28/ P. Brennecke, E. Warnecke, Waste Acceptance Requirements: Procedure and Basic Data, in: Merz, E., Odoj, R., Warnecke, E. (eds.), Proc. Int. Sem. on Radioactive Waste Products - Suitability for Final Disposal / Jülich 10. - 13.06.85, Berichte der Kernforschungsanlage Jülich, Jül-conf. 54, 1985, 497.

/29/ P. Brennecke, J. Schumacher, E. Warnecke; Radioactive Wastes with Negligible Heat Generation Suitable for Disposal, Kerntechnik 51, 1987, No. 2, 98.

/30/ Physikalisch-Technische Bundesanstalt; Plan Endlager für radioaktive Abfälle, Schachtanlage Konrad; Kurzfassung; Stand: September 1986.

Status of the spent LWR fuel disposal activities in the Federal Republic of Germany

K.-D. Closs

SUMMARY

As stipulated by the German Atomic Energy Act, reprocessing is the reference waste management route for LWR in the Federal Republic of Germany (FRG). Spent fuel disposal is being developed in addition to the realization of the reference concept. The aim of the ongoing program on spent fuel disposal is that the results will be available early enough to be incorporated in the licensing procedure of the Gorleben repository which will start as soon as the reults of the underground investigations show that the salt dome will be suitable for disposal.

Dr. K.-D. Closs, Kernforschungszentrum Karlsruhe, Projektgruppe Andere Entsorgungstechniken, Postfach 36 40, D-7500 Karlsruhe 1, Federal Republic of Germany

1 INTRODUCTION

Reprocessing is the reference waste management route for LWR in the Federal Repbulic of Germany (FRG). The basis for the German waste management policy is the German Atomic Energy Act of 1959. This law binds all participants in the nuclear fuel cycle

- to safely recover and recycle the resources contained in spent fuel or
- to classify the material as radioactive waste and to dispose it of safely, only if recycling is either scientifically or technically not feasible or economically not justifiable.

Large progress has been made in the FRG during the last years in closing the nuclear fuel cycle via reprocessing and recycling, as can be seen from other presentations at this seminar.

Spent fuel disposal without reprocessing has been under investigation in the FRG since 1979. The basis for those investigations was the waste management policy decision of 1979 agreed upon by the Heads of both the Federal Government and the States. This dicision stated that in addition to the realization of the reprocessing fuel cycle alternative back-end fuel cycle concepts, such as direct disposal of spent fuel without reprocessing, were to be investigated enabling a judgement by 1985, whether or not direct disposal might offer decisive advantages over reprocessing.

2 R&D PROGRAM "ALTERNATIVE SPENT FUEL MANAGEMENT AND DISPOSAL TECHNIQUES"

Between 1981 and 1984, the Federal Ministry of Research and Technology (BMFT) supported the R&D program "Alternative Spent Fuel Management and Disposal Techniques", which was coordinated by the Karlsruhe Nuclear Research Center (KfK). Within this program, Nukem and DWK developed the concept of a production scale conditioning and encapsulation plant and designed casks for spent fuel disposal. DBE (Deutsche Gesellschaft zum Bau und Betrieb von Endlagern für Abfallstoffe mbH, Peine), the company which will construct and operate waste repositories on behalf of the federal agency PTB

(Physikalisch-Technische Bundesanstalt, Braunschweig) designed the repository for spent fuel.

Within this program, the technical feasibility of spent fuel disposal has been demonstrated in principle. In addition, a comprehensive comparison of the two waste management options with and without reprocessing has been performed resulting in the following findings: As far as the status of technology is concerned, reprocessing constitutes proven technology in the FRG, whereas spent fuel conditioning and encapsulation is in the development stage. With respect to radiological safety, the comparison revealed that the radiological impact of direct disposal is generally somewhat lower than for reprocessing, but the difference is judged minor when discussing the absolute values and comparing them to natural radiation exposure. Cost of direct disposal of spent fuel seem to be lower than in the case of the reprocessing route. Finally, with regard to safeguarding a repository for spent fuel, some unresolved issues exist presently. It was concluded that direct disposal of spent fuel has not significant advantages over reprocessing /1, 2/.

The results of this evaluation formed the basis for a decision the Federal Government made in January 1985 as to how to proceed with direct disposal. It was stated that spent fuel disposal without reprocessing should be developed to technical maturity for those fuel elements for which reprocessing is either technically not feasible or economically not justifiable. That concerns, first of all, the thorium fuel elements from the German gas-cooled pebble-bed reactors, and, moreover, LWR fuel which is not within the scope of the Wackersdorf reprocessing plant specifications.

The activities already initiated or planned with respect to the development of spent LWR fuel disposal to technical maturity will be described in the next two sections.

3 TECHNICAL CONCEPT FOR SPENT FUEL CONDITIONING AND ENCAPSULATION

Within the R&D program "Alternative Spent Fuel Management and Disposal Techniques" the reference concept for spent fuel disposal was the emplacement of large and heavily shielded disposal casks in drifts of a repository mine located in a salt dome (Fig. 1). The main characteristics of this concept were as follows: Three intact fuel elements were sealed in a gastight canister, the dry storage bin. The disposal canister, surrounding the dry storage bin, consisted of a canister body and a corrosion protection which guaranteed an absolute confinement of the spent fuel for some hundred years. The lost shielding was designed such as to allow the workers to handle the packages hands-on in the repository.

In the meantime, cask development work was continued by DWK on their own, and a very flexible system with improved economics was designed dubbed the POLLUX Cask System /3/. This system can be used for all kinds of fuel elements. In addition to disposal, the cask system allows also for transport in the repository and long-term interim storage with an option of reopening the cask in case reprocessing is deemed desirable.

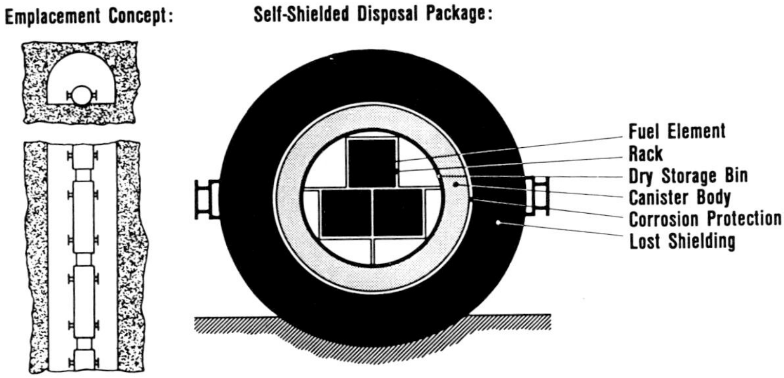

Fig. 1: Former reference concept for spent fuel disposal

Contrary to the old design, the POLLUX holds the rods from 8 spent fuel assemblies (Fig.2). The internals of the canister are divided into four equal quadrants and a void space in the center. Each quadrant is large enough to accommodate the rods from two PWR fuel asemblies while the compacted skeleton of the fuel assemblies is placed in the center void. The cask is designed to accept spent fuel three years after discharge from the reactor with a maximum heat load of 20 kW. The dimensions are similar to those of the old design, only the total weight is higher by a margin of about 20 % and totals 64 Mg. Some more characteristic data are shown in Table 1.

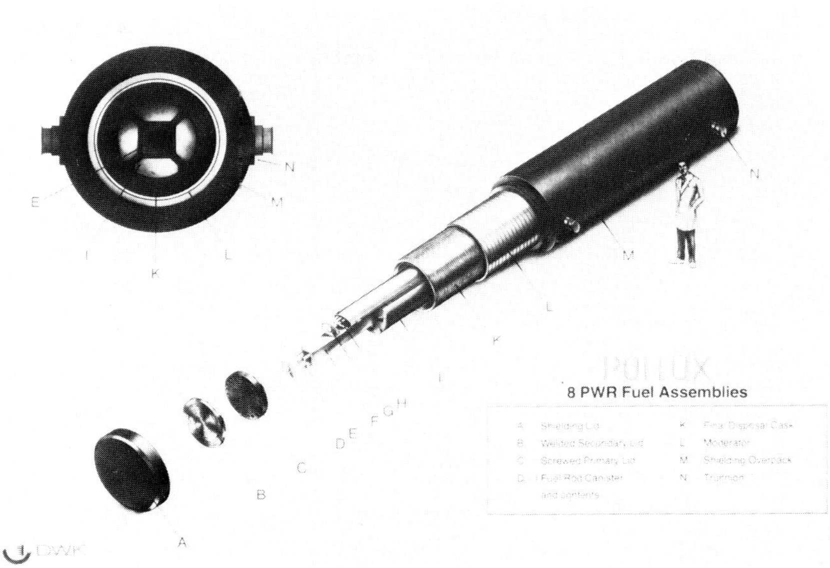

Fig. 2: POLLUX fuel disposal canister

The POLLUX containment concept is based on a double shell concept, defined by the disposal cask and the shielding overpack. The disposal cask consisting of a gastight welded steel containment assures the safe containment of spent fuel, the protection against mechanical loads and impacts and the protection against corrosion, which is realized by a Hasteloy C4 resurfacing welding process.

	Spent Fuel Package		HLW
	Large	Small	Package
Length (m)	5.35	1.33	1.33
Diameter (m)	1.57	0.43	0.43
Weight (Mg)	64	1.2	0.5
Number of fuel assemblies	8	0.5	2.2
Heat load (kW)[a]	6.5	0.4	2.0
Surface dose rate (mSv/h)[a]	<0.2	$<10^4$	10^6
Emplacement	drift	borehole	borehole

[a] 10 years after discharge from LWR

Table 1: Characteristics of LWR disposal packages

The shielding overpack assures, in addition to the radiation shielding function, further mechanical protection.

Besides this reference concept with large and heavily shielded packages, a back-up concept is being pursued which results in smaller spent fuel canisters. In this case the disassembled fuel pins have to be cut into pieces of about 1 m length. As can be seen from Table 1, the small spent fuel packages have the same outer dimensions as the canisters for vitrified reprocessing waste. The advantage is that the same handling and emplacement technique can be used in the repository as for vitrified waste, namely the emplacement of the canisters into 300 m deep boreholes, drilled in the floor of the disposal galleries.

In May 1986, DWK initiated the licensing procedure for a pilot conditioning and encapsulation plant /4/ enabling the encapsulation of all kinds of spent nuclear fuel in different disposal casks. The POLLUX system described above is used as the reference concept. This plant will be built at Gorleben at the site of the interim storage facility. According to the present time schedule, the cold and hot operation could start in 1992 and 1994, respectively.

4 REPOSITORY RELATED R&D

Contrary to spent fuel conditioning and encapsulation, the repository related R&D will be further supported by the BMFT. The R&D program "Direct Disposal of Spent LWR Fuel" is coordinated again by KfK. Major contributions to the programm are made by DBE, GSF (Gesellschaft für Strahlen- und Umweltforschung, Institut für Tieflagerung, Braunschweig), BGR (Bundesanstalt für Geowissenschaften und Rohstoffe, Hannover) and KfK.

Up to now, the planning for a repository in the Gorleben salt dome was concentrated on accommodating reprocessing waste but it did not take into account the emplacement of large and heavy spent fuel packages. Even a small number of spent LWR fuel elements intended for direct disposal could entail major changes in both the aboveground and underground design of the repository. With respect to the reference concept, the mining authorities have pointed out that the feasibility of direct disposal is to be demonstrated before a license for industrial scale deployment could be granted.

Thus, the main thrust of the program lies with the demonstration of the safe and reliable handling of large and heavy canisters in a repository on a 1:1 scale. Emphasis is also laid on the investigation of special thermal and rock mechanics aspects of the drift emplacement concept as well as on the layout and optimization of a repository for both reprocessing waste and conditioned spent fuel.

Figure 3 shows the logistics for POLLUX casks in the repository. In the repository shaft area, the cask is loaded on a rail transport vehicle. The loaded vehicle is slid into the cage and then lowered to the emplacement level. Since the shaft has a diameter of 7.5 m, the 5.4-m long POLLUX cask can be transported in the cage in a horizontal position. Railbound transport takes place as far as to the emplacement drift. A hoisting apparatus is installed at the position where the cask is to be disposed of. After the transport vehicle has reached this position, the cask is lifted a few centimeter, the vehicle is pulled away, and the cask is lowered on to the floor of the drift. Soon afterwards, this part of the drift is backfilled with crushed salt.

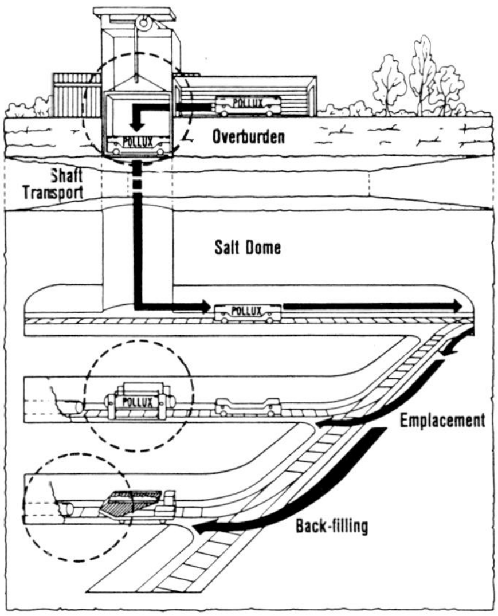

Fig. 3: Logistics for POLLUX casks

The three areas in which further R&D with regard to demonstration tests /5/ is necessary, namely shaft transport, underground handling and emplacement as well as the thermal and mechanical behavior of the back-filled drift, are marked on Fig. 3:

- The shaft transport with loading and unloading of the hoisting cage will be simulated in an aboveground test rig with inactive canisters. Up to now, shaft transport with a payload of 80 Mg (POLLUX + transport vehicle) has not yet been demonstrated, but existing technology can be adapted. The following topics will be investigated, for example: locking of the hoisting cage, pushing the loaded vehicle onto the cage, locking of the vehicle within the cage, loading of the hoisting cables and hoisting machinery upon release of loaded cage, locking of the loaded cage, unloading of the cage etc.

- The manipulation of inactive large disposal casks and the demonstration of safe handling will be done above and below ground. The machinery for the transport of the casks into the disposal drift as well as for lifting and lowering the canister at the disposal position is to be developed, constructed on a 1:1 scale, and tested. The objective is to obtain general approval for the components under the provisions of the mining law.

- Thermal and rock mechanics investigations for the drift emplacemnt concept will be performed with electrically heated canisters in the ASSE underground laboratory, as shown in Fig. 4. For this test, two parallel drifts are excavated, each of which is to accomodate three canisters with an electric power of 8 kW each. The casks are surrounded by crushed salt. Different types of measuring instruments are installed in the crushed salt and surrounding strata in order to record and evaluate all relevant parameters (temperature, pressure, roomclosure rate, heat flow, compaction of the crushed salt etc.) throughout the duration of the test which will last for some years.

Beside the aforementioned demonstration tests, the layout and optimization of all aboveground as well as subsurface process steps with regard to a common repository for both reprocessing waste and conditioned spent fuel is necessary /2/. Different mixes of spent fuel that is going to be disposed of directly and reprocessed, respectively, are the basis for this systems analysis "Dual Purpose Repository". Concept variations are generated by varying cooling times prior to disposal, by conceiving different conditioning and canister

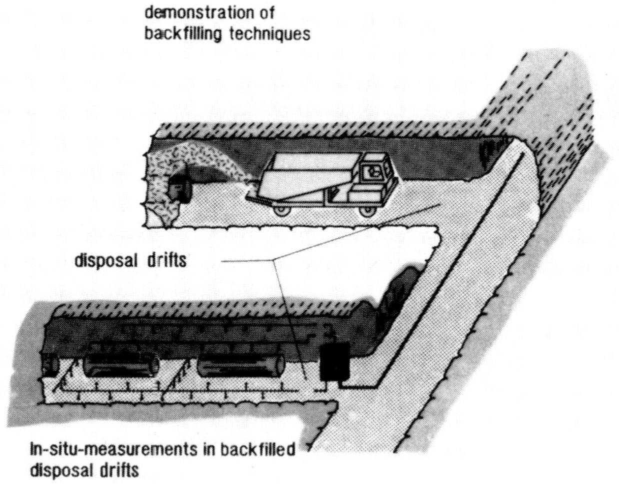

Fig. 4: Inactive thermal simulation of the emplacement in disposal drifts

concepts, and finally by taking into consideration a number of repository design alternatives and emplacement techniques, such as shown in Fig. 5. The set of repository alternatives will include either drift or vertical borehole emplacement as well as a combination of both concepts. Emplacement drifts and boreholes will be located either in different sectors of the model salt dome or in one common sector. Emplacement of heavy disposal casks will even be studied for drifts being arranged at different levels of the repository. Conceptual design of the various repository and emplacement concepts will be accompanied by geothermal and thermomechanical calculation. The pros and cons of the system variants will be determined by quantifying the performance criteria "expenditure until technical maturity", "radiological safety during routine operation", "long-term safety of the repository", and "cost". The final evaluation will also include the areal requirements of the disposal concepts and aspects of international safeguards.

5 CONCLUSIONS AND OUTLOOK

In addition to the realization of the fuel cycle based on reprocessing, spent fuel disposal without reprocessing has been investigated during the last years. After having proven the principal technical feasibility of this waste management route, the program has reached a new phase: During the next years, important aspects of this

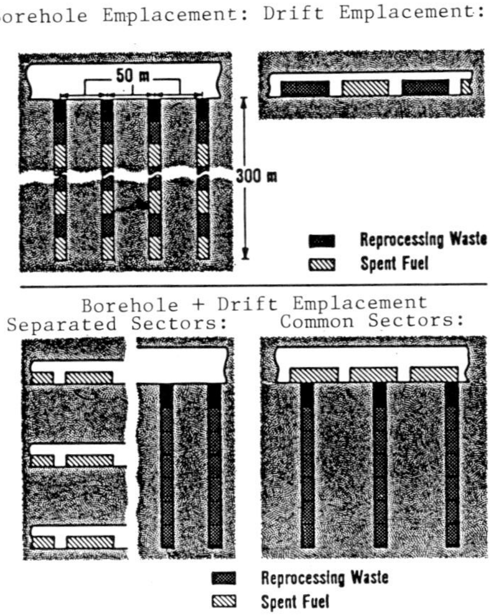

Fig. 5: Emplacement concepts

back-end fuel cycle concept will be demonstrated. For this reason, a pilot conditioning and encapsulation plant will be performed on a 1:1 scale. The aim of the ongoing program on spent fuel disposal is that the results will be available early enough to be incorporated in the licensing procedure of the Gorleben repository which will start as soon as the results of the underground investigations show that the salt dome will be suitable for disposal. This means that -in addition to the reprocessing and recycling waste management concept- direct disposal of spent fuel not suitable for reprocessing could be introduced as an additive concept for completely closing the backend of the fuel cycle.

LITERATURE

/1/ Systems Study Alternative Entsorgung. Executive Summary. Karlsruhe Nuclear Research Center (1985)

/2/ PAPP, R. and CLOSS, K. D., "Results of the German Alternative Fuel Cycle Evaluation and Further Efforts Geared Toward Demonstration of Direct Disposal", Proc. Waste Management '86 Symposium, Tucson, Arizona, March 1986, Volume 2, p. 523-526

/3/ EINFELD, K. and POPP, F. W., "Dry Storage Systems Using Casks for Long Term Storage in an AFR and Repository", Third International Spent Fuel Storage Technology Symposium/ Workshop, Seattle, USA, April 1986

/4/ EINFELD, K., "Pilot Plant for Spent Fuel Conditioning at Gorleben", Nuclear Europe 3-4/1987, p. 21

/5/ ENGELMANN, H.-J., HARTJE, B., SCHRIMPF, C., "Demonstration of the Spent Fuel Disposal Techniques Test Program in the Federal Republic of Germany", Proc. Waste Management '86 Symposium, Tucson, Arizona, March 1986. Volume 2, p. 531-534

Long-term safety assessment for the disposal of nuclear waste

E. R. Merz

SUMMARY

A matter of concern amongst experts is the length of time for which safety analysis of nuclear waste repositories should be performed. Although time perspectives between 10^3 and 10^7 years are under discussion, the persuasion is growing that a time frame of about 10,000 years should be appropriate for calculating potential dose rates to individuals living in the vicinity of a disposal site. Beyond this time the uncertainties in forecasts and future assessments are greater the longer the periods of time they are applied to. It therefore becomes very questionable whether there is any use in continuing long-term assessments solely on the basis of risk to hypothetical individuals who constitute a critical group well beyond this time. It is therefore proposed to calculate individual dose rate exposures over a time period of 10,000 years based on realistic site-specific scenarios. No dose rate calculations regarding post 10.000 year releases are recommended, but qualitative forecasts concerning a safe isolation of waste should be made up to 10^6 years using best estimates of geological and hydrogeological data, especially based on ground water travel times. Two main arguments have been put forward to support this approach: (1) a comparison of the waste repository with natural analogues, and (2) the orientation to future geological alternations influencing the repository system. Questions of predictive modeling a nuclear waste repository system are briefly discussed stressing the importance of uncertainty evaluations in regard to the reliability of such an assessment obtained with the available data used.

Prof. Dr. E. R. Merz, Institute of Chemical Technology, Kernforschungsanlage Jülich GmbH, D-5170 Jülich, Federal Republic of Germany

1 INTRODUCTION

The objective of underground disposal is to isolate radioactive waste from the biosphere for a period of time such that a release of radionuclides from the repository will not result in dangerous radiation exposure. Decisive is the direct impact of radiation to man and only in an indirect way, the inventory of radioactivity confined in the repository, since a multibarrier containment system prevents almost any release of radionuclides to the human environment /1/. After the operational phase of a repository, that means once the repository has been sealed, a long-term radiological risk remains. Due to the radioactive decay, it decreases with time but must nevertheless be quantitativly assessed.

The long-term radiological impact of waste disposal depends on events and processes which may cause a release of radionuclides into the environment or influence the rate of release or transport through it. Some of these incidents are almost certain to occur, others have time-dependent probabilities of occurence.

In the absence of any possibility of direct demonstration of the long-term safety of particular geologic disposal options over the periods for which high radioactivity levels persist, the demonstration of long-term safety must be indirect. It consists of a theoretical assessment of the long-term safety, employing predictive modeling for the overall evaluation, but using experimental information for the physical phenomena and the associated data.

A major problem in these assessments are "uncertainties" which are basically of three types:

1. scenario uncertainties,
2. model uncertainties, and
3. parameter uncertainties.

Most important are the latter since they have the greatest influence on the statement.

2 ASESSMENT OF REPOSITORY PERFORMANCE

It is generally considered that, if radioactivity is released from buried waste, it is most likely to result from ground water entering the repository and gaining access to the waste form, dissolving some radionuclides and then transporting them to the surface. Other methods of release generally considered to be of much lower probability are the exposure of waste by some catastrophic natural event or through future mining operations. These may be largely avoided by locating waste repositories at sufficient depth in an area known to be seismically and volcanically stable, and in a geological environment unlikely to contain minerals of economic value.

The assessment of long-term waste confinement first requires an identification of the various potential mechanisms, rates and pathways by which radionuclides could leave a repository and migrate through the various barriers to enter the human enviromnment and bring about radiation exposure either directly or indirectly. Once these factors are identified, estimates can be made of the radiation doses that could arise through these mechanisms.

2.1 CLASSIFICATION OF GEOLOGICAL EVENTS

Possible events which could lead to repository failure can be divided into four groups:

1. natural rapid events uninfluenced by the existence of the repository or by human actions, e. g. meteorite impact, volcanic activity;

2. natural geological events or processes independent of repository existence, e. g. erosion, subsidence, faulting, climatic changes;

3. geological events or processes caused by the repository, e. g. thermal and radiation effects on rocks;

4. human actions, direct or indirect, e. g. drilling for mineral resources, reservoir construction, sabotage, war.

Table I summarizes the events which have been identified by HILL and GRIMWOOD /2/ as potential causes of failure of geological containment and the possible consequences leading to transport of waste to the biosphere. The inherent difficulties in performing long-term risk analyses stem from the lack of assigning probabilities to the individual events.

Scenario Mechanism Potential Cause of Containment Failure	Probability of Event Occurring	Possible Consequences
Rapid natural events		
Meteorite impact	$2 \cdot 10^{-14} y^{-1}$	Direct release to atmosphere and/or leaching into ground or surface water
Volcanic activity	$5 \cdot 10^{-11} y^{-1}$	Direct release to atmosphere, hydrothermal transport in deep ground water
Slow natural processes		
Seismic events and faulting	$1 \cdot 10^{-5} y^{-1}$	Ground water ingress
Erosion	1	Increased vulnerability to other failure events and transport mechanisms
Climatic changes with glaciation	Almost certain in less than $2 \cdot 10^{4} y$	Reduction in rock strength, hydrological changes leading to ground water ingress, also increased erosion
Pluvial episodes	Unknown	Possible increase in probability of ground water ingress
Arid episodes	Unknown	Probably none
Uplift	Unknown	Increased erosion, hydrological changes leading to ground water ingress
Subsidence	Unknown	Probably none
Repository- and waste-induced effects		
Thermal effects	Unknown	Fracturing allowing ground water ingress
Radiation effects	Low	Fracturing allowing ground water ingress
Mechanical stresses	Unknown	Fracturing allowing water/solution ingress
Events caused by man		
Nuclear warfare	Unknown	None expected
Sabotage	Unknown	Probably none
Drilling	Unknown	Small-scale release to surface

Tab. 1: Summary of geological containment failure mechanisms

2.2 COMPLEXITY OF A REPOSITORY SAFETY ANALYSIS

An overall safety analysis for a repository not only requires the identification of geological containment failure mechanisms but a complex array of geological, hydrological and geochemical data. Much of this can be provided by the existing general background of knowledge of these sciences. However, equally important data required for the analyses can be obtained only from detailed knowledge of the properties of the proposed disposal sites. This requires an extensive preliminary program of testing and evaluation of repository sites.

The principal factors governing the overall performance of a repository are depicted in Table 2.

All of these factors must be assessed on the basis of the best available evidence, including realistic estimates of their uncertainties. When this data base is assembled, calculations of the release of radionuclides to the human environment may be carried out using mathematical models which simulate the overall system. These models have been of two types: consequence analyses and scenario analyses /3/.

Hydrological environment (in particular, the routes of movement of ground water through the repository and its neighborhood)
Waste form and engineered barriers
Oxygen potential of the rocks and ground water or brine both within the repository and in the surrounding rocks
Composition of the ground water and brine
Solubilities of radionuclides in ground water and brine
Time scales for movement of water and brine from the repository to the biosphere
Retardation factors for individual radionuclides due to adsorption-desorption phenomena on clays and other minerals as water and brine solutions diffuse along various pathways to the biosphere
Degree of dilution of ground water and brine solutions emerging from the repository with non-contaminated ground water from environments away from the repository

Tab. 2: Processes and factors governing the overall performance of a repository

Figure 1 shows the components of safety analysis and the techniques which may be used at each stage.

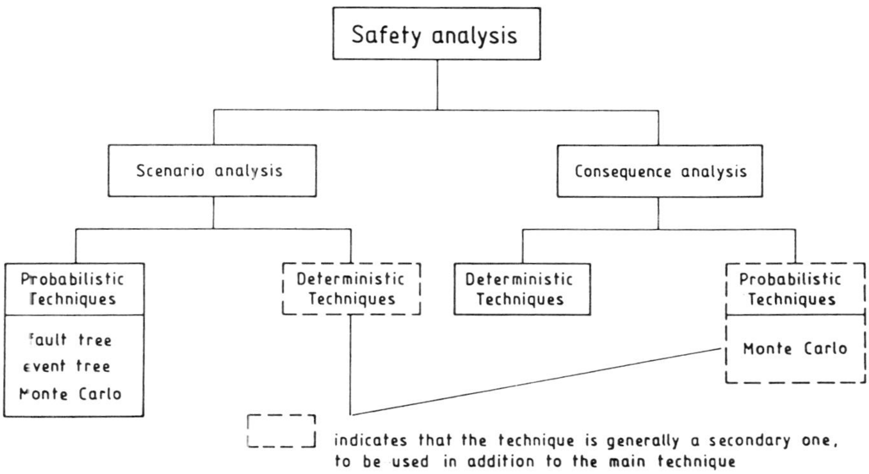

Fig. 1: Components and techniques of safety analysis

2.3 DETERMINISTIC ANALYSIS TECHNIQUES

Deterministic analysis techniques are the classical methods used in predictive mathematical modeling of system behavior. In order to use these techniques it is necessary to have sufficient understanding of the processes at work on and within a system to be able to formulate the mathematical equations describing the principal processes. Detailed knowledge of each process is not necessarily required, since it will often be possible to develop a mathematical model which adequately predicts the behavior of a system or system component from a general understanding of the basic processes involved. Deterministic techniques are particularly useful in modeling the effects of continuous processes and can be applied to both steady state and dynamic conditions which change with time.

2.4 PROBABILISTIC ANALYSIS TECHNIQUES

Probabilistic analysis involves a set of statistical techniques for studying effects. Parameters whose values are uncertain, events whose occurrence are random, and features which may or may not be present can be treated statistically. For example, the frequency or probability of occurrence of an airplane crash at a specific repository site may be estimated. Various methods are available for considering how the probabilistic variations in components of a system act together to cause variation in the system as a whole. These include fault/event tree analysis and Monte Carlo analysis.

It is important to recognize that probabilistic analysis and deterministic analysis are complementary techniques and that, as a general rule, both should be used in scenario as well as in consequence analysis.

Consequence analyses are primarily determinstic in nature, however, when distribution functions of environmental transport parameters are considered, consequence analyses also have probabilistic components. If sub-models of particular system components are used, these must be able to be readily interfaced to produce a model of the whole disposal system.

A scenario analysis is necessarily probabilistic because the probabilities of occurrence of various events and processes have to be taken into account. However, it also has deterministic components, since sub-models will need to be used to predict the effects of various events and processes on the disposal system and hence to provide the input data for release and transport calculations.

3 LESSONS TO BE LEARNED CONCERNING THE TIME PERIOD

From the outcome of numerous risk assessments of radioactive waste disposal systems, it may be concluded that the risks identified over relatively short time span, such as a few hundred to one thousend years, do not adequately portray important differences between alternative sites or disposal systems. This is because the ground water travel times would probably be sufficiently long at most sites so that no significant radionuclides releases would be predicted over this time period.

If the consequence analysis technique is carried out further into the future using a time frame up to about 10,000 years, then in most cases substantial differences between different sites could be observed. Underlying chemical, physical and geological processes are either known from experiments and observations, or can be predicted with sufficient confidence for this time period. Beyond 10,000 years the uncertainties in forecasts and future assessments become greater the longer the periods of time they are applied to. Applying probabilistic instead of deterministic techniques does not change this situation in the least. Therefore, it is recommended that safety analyses be divided into two categories:

1. A time frame of up to 10^4 years for which the consequence analysis technique provides reliable quantitative results in the form of release data and radiation exposure values.

2. Beyond 10^4 years up to the maximum of 10^6 years where the safety analyses are performed only in a qualitative manner without any exposure calculations since they would become highly speculative. An absolute time cut-off is applied for time frames greater than 10^6 years.

Although this selection is largely arbitrary, the adoption of the 10,000 years limiting value seems to be a scientifically justifiable approach.

4 SUPPORT FOR THE TIME FRAME ASCERTAINMENT

The selection of the 10,000 years limiting value for making quantitative safety analyses by calculating radiation doses to man from radionuclide release and transport phenomena calls for supporting arguments. A desirable method would be a comparison of the hazard potential and risk, respectively, with a natural analogue. Thus, in the past a toxicity index has been derived in order to compare it with natural standards /3/.

The purpose of a toxicity index is to allow a simple comparison of the toxicity of a repository with known standards, without having to take account of all the processes and mechanisms which control the release and migration of radionuclides from the repository to the biosphere. It is important to point out that although radiotoxicity is a measure for hazard potential, it does not display the actual risk which is additionally dependent on the retardation efficacy of the multibarriers.

Two types of comparison are generally made: with natural uranium bodies and with toxic heavy metal deposits. An additional argumentation for a cut-off limit is based on drastic climatic changes with glacial and interglacial episodes.

4.1 RISK FROM URANIUM ORE BODIES AND MILL TAILINGS

A general criterion has been suggested for deep geological repositories containing radioactive waste in the sense that the repositories should impose no greater radiological risk than that due to naturally occurring uranium deposits. Although the true hazards of radioactive wastes are not measured by comparing the total ingestion toxicity index for the uranium ore used to fuel the reactor to generate these wastes. In Figure 2 the toxicity indices are shown relative to the ingestion toxicity of the ore.

The ore toxicity is due mainly to Ra-226, which is in secular equilibrium. Also shown are the relative toxicity indices for uranium mill tailings, which contain Th-230 and Ra-226 separated from the uranium ore, and for depleted uranium from isotope separation, neglecting the likely later use of this uranium as fuel for breeder reactors. Because the uranium ore ingestion toxicity is dominated by

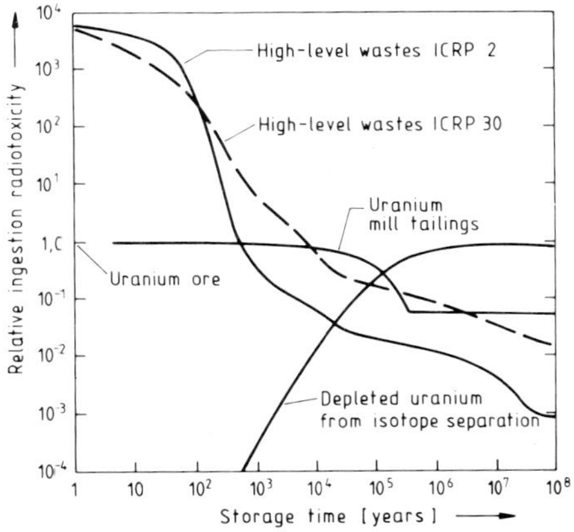

Fig. 2: Relative ingestion toxicity of solid residuals from LWR fuel cycle (uranium fuel, 0.5 % of U and Pu in high-level waste)

Ra-226, all of this toxicity is transferred to the mill tailings and is preserved for over 100,000 years because of the long half-life of Th-230. The tailings toxicity then decays to a lower value due to the residual uranium, e. g. about 5 %, which remains with the mill tailings.

The decrease of the ingestion toxicity of the high-level waste is dependent on the gross decay rate of fission products and the conversion factors applied. In the diagram the values issued by ICRP recommendations No. 2 and the lastest one No. 30 are used to calculate the relevant curves. With the former data the high-level waste decays to a level below that of the initial ore after less than 1000 years, whereas with the more recent data the point of intersection lies close to 10,000 years. This difference is due to new metabolic data as well as new radiobiological knowledge, and as a consequence thereof, modifications for exposure calculations via the ingestion path. It may thus be argued that after about 10,000 years there no longer exists a net increase in radiotoxicity potential.

Another perspective to justify the proposed time period is based upon a comparison of the risks from a high-level waste repository with those from an undisturbed uranium ore body. In this assessment, it is assumed that uranium dissolved by the ground water eventually reaches an existing water path. The uranium itself and some of its daughter products move along the aquifer and finally may enter into the biosphere. The same progapation model is applied for the mobilized waste radionuclides. Figure 3 compares the toxicity of uranium ore required for the production of fuel with the relevant waste arising from reprocessing, and the spent fuel itself.

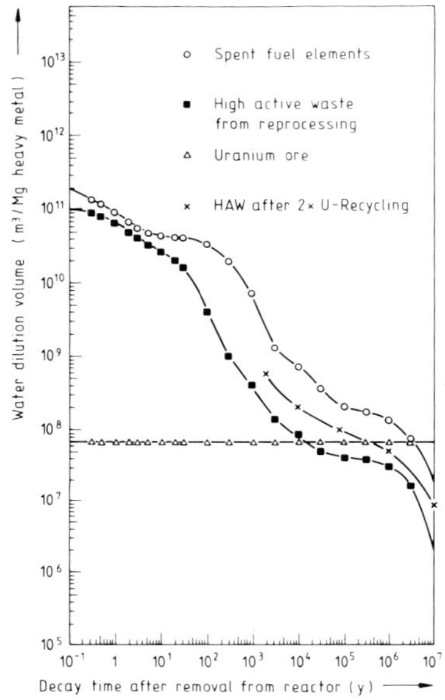

Fig. 3: Toxicity of nuclear wastes relative to the amount of uranium ore (0.2 % U content) required for the production of fuel (adapted from HAUG) /4/

4.2 RISKS FROM HAZARDOUS MINERALS AND HEAVY METAL DEPOSITS

Radioactive wastes are occasionally compared with other non-radioactive hazardous materials like ore bodies, fly ashes and incineration slags, or with any dumping site containing chemotoxic materials, e. g. compounds such as cyanide, arsenic, mercury, cadmium, lead. Calculating the toxicity index for such buried waste, assuming a uniform distribution over the areal extent of the underground repository, and comparing it with an equal size nuclear waste repository, it can be shown that they exhibit rather similar toxicities after periods form 10^3 to 10^5 years decay time for the radioacive material /5/.

Another approach compares radioactive wastes from nuclear power production with the wastes arising from coal burning. The results of a recent study by the PTB in Germany are shown in Figure 4 /6/.

The principal limitation of such a comparison is obvious since many important factors of a disposal system are omitted from the analysis. The mobilization, pathways and dispersion of hazardous species between the repository and the local human environment may be rather different, although the toxicity contents in the disposal areas are of comparable dimensions.

A more important objection to such a vindication is the fact that it yields no independent absolute measure of a hazard or risk.

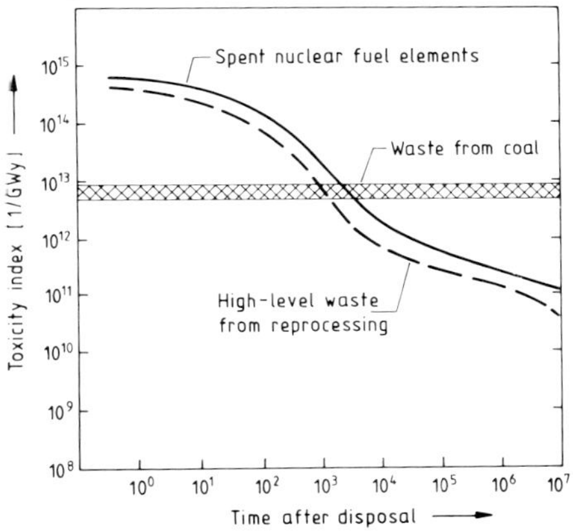

Fig. 4: Relative toxicity indices as a function of time for wastes arising from nuclear power production and wastes from coal-fired plants which are released to surface waters and thus enter into the human biosphere

4.3 RISKS FROM CLIMATIC CHANGES

Phenomena which over the ten thousend years of interest may change important barrier parameters are events with time-dependent probabilities of occurrence that change the behavior of a disposal system or that give rise to new exposure pathways. Very few such events are random in time, and more geological events of potential influence on a disposal facility, such as changes in sea levels or glaciation, appear to have a predictable periodicity /7/.

A model of future climate based on observed orbital-climate relationships predicts that, ignoring anthropogenic effects, the long-term trend over the next 10.000 to 20.000 years is towards extensive glaciation of the northern hemisphere and a cooler climate. It is therefore extremely probable that the repository sites in Germany will be affected within this time frame.

Subsequent effects may influence geological containment. Perhaps the most obvious one is increased erosion. Another certain consequence is that glaciation causes great changes in drainage patterns, but predictions of these hydrological changes are difficult. Figure 5 gives an overview of the geological time perspective in the past and for the future.

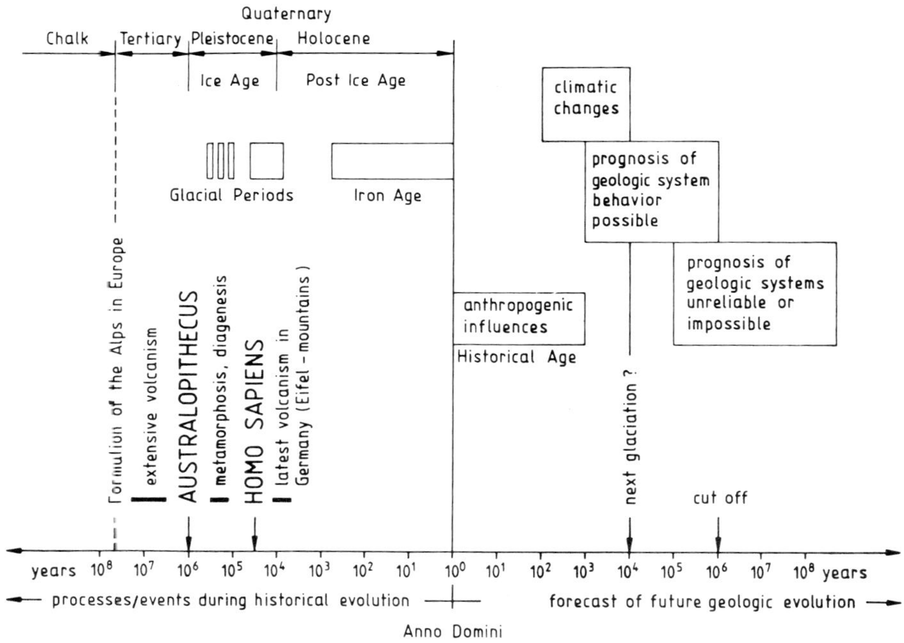

Fig. 5: Schematic diagram of geological perspective

5 IMPROVEMENT OF SITE-SPECIFIC SAFETY ANALYSES

A site-specific safety assessment is a licensing prerequisite. The scenario uncertainties are narrowed down by the general consensus that a transport medium, e. g. ground water or brine solution, is required that has to come into contact with the waste. An assumed intrusion of aqueous solution represents an "envelope" scenario because it might result from a variety of causes such as fault formation and growth, salt dissolution by groundwater, thermally induced fractures, bore hole and mine drift seal failures or brine migration.

The calculation of the behavior of a repository requires the mathematical description of experimentally based physical models. The quantification of the direct modeling uncertainty is difficult. However, parts if not most of the modeling uncertainties may be accounted for by added parameter uncertainties.

There are large uncertainties associated with many of the input parameters to the models. The quantification of these parameter uncertainties may vary from the specification of only a single mean value, or it may consist of upper and lower bonds, or of an entire probability distribution for certain input variables.

An approach to improve the unsatisfactory situation has been proposed by PEREZ-RAMIREZ and OTT /8/ by virtue of including a probabilistic evaluation of the effects of parameter uncertainties. It results in average release rates based on the probability distribution assigned to the variables controlling the waste migration for a failure scenario which assumes the intrusion of ground water into the repository.

In this approach "transfer kernels" are applied for each individual medium. The superposition of these kernels describe the overall migration process. The probability distribution of parameter uncertainties are integrated with the deterministic kernel K (τ) to yield the average transmission rate, K (τ), between interfaces.

A simplified two-region model is treated as a sample problem. The first "region" represents the underground facilities of a saltdome repository and the second region represents an aquifer, described by a one-dimensional model. As a sample problem, two long-lived radionuclides have been treated so far: I-129 and Np-237. The specific accident scenario and the selection of data have been adapted from the German "Projekt Sicherheitsstudien Entsorgung", PSE /9/.

An example of release rates from the repository for Np-237 are presented in Figure 6.

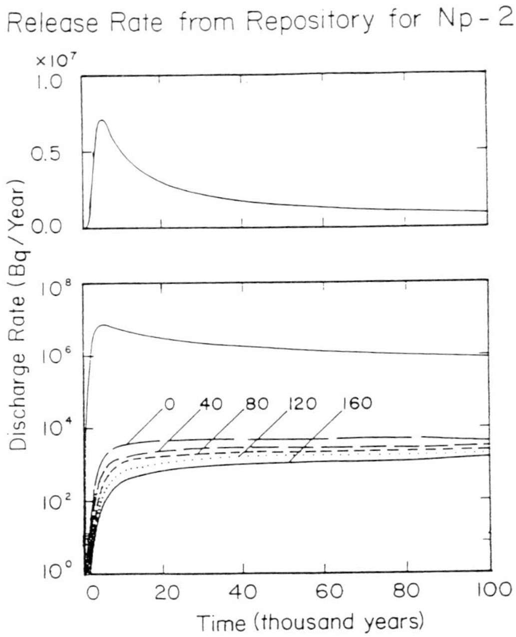

Fig. 6: Average Release rate from repository for Np-237. The results are depicted for five different brine intrusion times (in years). The average release rates for Np-237 start increasing more showly than the calculated PSE release rate for the first few several thousand years as can be deduced from the diagram. In general, the average release rates calculated by PSE are higher by one to two orders of magnitude.

6 CONCLUSIONS

A justification for the selection of the 10,000 years time period on the basis of a comparion with other hazard potentials, e. g. uranium ore bodies, toxic heavy metal containing ashes and slags from fossil-fired power plants is of a qualitative nature only. It provides by itself no quantitative measure of a risk.

However, its usage supports public acceptance of nuclear power utilization.

A direct justification for the 10,000 years time period is based upon geological reasoning. Forecasts of future developments are only possible for this time span with sufficient reliability.

Reliable quantitative data can only be derived from a site-specific safety analysis comprising the integrated system: "overall geological situation-repository layout-waste package".

A successive procedure is recommended:

- Quantitative radiation exposure calculations via release and dispersion data up to a time frame of 10,000 years. Protection goal: individual dose limit < 0.3 mSv/yr, no risk upper bound commitment.

- Best estimate considerations about the geological, hydrogeological and geochemical behavior of the multibarrier system for the time period between 10^4 and 10^6 years. Only release rate, but no radiation exposure estimates.

- Cut-off at 10^6 years since the residual risk is equal to or less than the overall general geological underground hazard potential.

LITERATURE

/1/ E. Merz, K. Kühn, M. Bloser, A. Matting: "Implementation of the Federal Republic of Germany's Policy and Strategy of Radioactive Waste Disposal". Proc. Int. Symp. on the Siting, Design and Construction of Underground Repositories for Radioactive Wastes, Hannover, FRG, 3-7 March 1986, IAEA-SM-289/9

/2/ M. D. Hill, P. D. Grimwood:
"Preliminary Assessment of the Radiological Protection Aspects of Disposal of High-Level Waste in Geologic Formations". National Radiological Protection Board, Report NRPB-R.69, January 1978, Harwell, Didcot, U. K.

/3/ J. Hamstra:
"Radiotoxic hazard measure for buried solid radioactive waste". Nuclear Safety 16 (1975) 180-189

/4/ H. O. Haug:
"Management of Radioactive Wastes from the Nuclear Fuel Cycle". Proc. of a IAEA Symposium, International Atomic Energy Agency, Vienna, Vol. 2 (1976) 233

/5/ C. M. Koplik, M. F. Kaplan, E. Ross:
"The safety of repository for highly radioactive wastes". Review of Modern Physics 54 (1982) 269-310

/6/ D. Ehrlich, H. Rothemeyer, G. Stier-Friedland, E. Thomauske:
"Langzeitsicherheit von Endlagern" Atomwirtschaft/Atomtechnik 31 (1986) 231-236

/7/ H. Flohn:
"Eiszeit oder Warmzeit? Fakten und Überlegungen zur Klimaentwicklung". Naturwissenschaften 66 (1979) 325-330

/8/ G. A. Perez-Ramirez, K. O. Ott:
Probabilistic Safety Analysis for the Disposal of Nuclear Waste. Doctoral Thesis, School of Nuclear Engineering, Purdue University, West Lafayette, India/USA, June 1987

/9/ Projekt Sicherheitsstudien Entsorgung, PSE:
Zusammenfassender Abschlußbericht, Hahn-Meitner-Institut Berlin, HMI, Januar 1985